Letitia

She's Iron, Stone, and Beautiful Flowers

Colorado Homesteading 1916 to 1950

By Barbara Ann Black

To Cheena with love

Barbara Black

Naturegraph Publishers Inc.

Library of Congress Cataloging-in-Publication Data

Names: Black, Barbara, 1928- author.
Title: Letitia : she's iron, stone, and beautiful flowers : Colorado homesteading, 1916 to 1950 / by Barbara Ann Black.
Other titles: Colorado homesteading
Description: First edition. | Happy Camp, California : Naturegraph Publishers, Inc., [2018]
Identifiers: LCCN 2018044205 | ISBN 9780879612900
Subjects: LCSH: Black, Letitia Brunia, 1889-1974--Family. | Black, Barbara, 1928---Family. | Black family. | Women pioneers--Colorado--Biography. | Frontier and pioneer life--Colorado. | Ranch life--Colorado--History. | Ranchers--Colorado--Biography.
Classification: LCC F781 .B43 2018 | DDC 978.8/03092 [B] --dc23
LC record available at https://lccn.loc.gov/2018044205

ISBN: 978-087961-290-0

2018 Printing

Printed in USA

Naturegraph Publishers has been publishing books on natural history, Native Americans, and outdoor subjects since 1946. Free catalog available.

Books for a better world

Naturegraph Publishers, Inc.
PO Box 1047 • 3543 Indian Creek Rd.
Happy Camp, CA 96039
(530) 493-5353
www.naturegraph.com

Dedication

To my wise, wonderful mother,
Letitia Brunia Black,
her beloved grandchildren,
Tamara, Roxana and Keven,
and their children, who
will carry forward her lineage.
The 3N ranch is left to posterity,
dedicated to keeping our "roots"
intact for Letitia's descendants.

Acknowledgments

Chester Kahn, Navajo artist, painted the cover and drew the illustrations. Chester, while tending sheep as a boy, drew pictures on the canyon walls. I also herded sheep as a child, bringing them to our canyon while watching swallows overhead tending mud nests.

Jeffery Black, my cousin, two months older, shared in some of my experiences. He recounted additional details about some of the stories and names in this book. Having no brothers, sisters, or neighborhood children nearby, when he visited the ranch he was like a brother to me as we played or worked together.

Son, Keven Brown, enjoyed each newly written ranch story I shared with him and insisted that I continue to write this book. As my editor, he has been instrumental in improving its organization, fact checking details, and making the stories more readable.

Grandson, Travis Pinn, created the ranch maps and helped to arrange the subject matter and table of contents.

Granddaughter, Minea Good, helped research the ranch birds.

Albert Vanderhoof (Sonny) undertook the challenge of reproducing old photos from Mother's photo album, many taken a hundred years ago, and also did the layout and typesetting of the book. He photographed the ranch keepsakes for pages, 17, 48, 50, 88, 114, 122, 218, 220, 248, 261 and 262.

All photos in this book are from family photo albums or taken by the author unless acknowledged under the photo, or are family keepsakes shown on pages mentioned above and photographed by Albert Vanderhoof.

Many who inspired me to write about Mother's life over the past forty-some years sadly have passed on. I thank my family and friends for decades of persistently keeping after me, with persuasion and encouragement, to see this book through to publication.

Table of Contents

Letitia Brunia , ca. 1908

Prologue

It has been said that when a person dies, a library of knowledge is lost. I do not wish that to happen with my mother, Letitia Brunia Black (b. May 17, 1889). A small portion of her practical wisdom is captured in this book, as illustrated in the accounts of how she and her two sisters started and built their homesteads in 1916 though the 1920s, and how she and her husband (including a young daughter now) started again from scratch when they returned to Colorado at the beginning of the Great Depression. Her ingenuity to overcome the numerous obstacles they faced (including being broke, the long drought, the grasshopper siege, and the Dust Bowl years of the 1930s), her knowledge of frugality, and her pioneer skills that served her family well in the years ahead are lessons as valuable today as they were then.

In conversations with and letters from neighbors and friends, I've only heard compliments about Mother. A nurse friend from Iowa wrote: "She was one of the gentlest, kindest persons I ever knew.... I classed her with my mother—they both had great tolerance and compassion....She could do anything with her hands...to me she was a genius." A homesteader's daughter wrote: "She was one in a million. Anyone that knew her had their lives enriched. Some of my fondest memories were [of time] spent on her homestead."

Mother and Dad had been away from the homestead for over three years when the Great Depression hit hard and banks closed across the United States, hastening their return to Colorado. It was 1931 and I was almost three. Roads to and from our ranch then were dirt "wagon roads", thirty-five miles to Fowler or thirty-five miles to Walsenburg, the two nearest towns, and ten miles to the mail boxes and our closest neighbor with a phone. The homestead had no electricity or running water, just mother's two-room, sod-roofed cabin, a windmill, an old gray barn, and the land. It was essential for us to never get sick or hurt. For my safety, I was to mind at once. One day on the prairie, Mother gave me the perfect example. We saw a mother quail and her covey of twelve tiny babies. Just then a hawk flew overhead and mama quail warned her offspring to hide. Instantly, they disappeared so well, even with scant grass cover, that although I had seen them vanish, not one could I find. When the hawk was out of sight, mother quail gave a signal that they could come out. Mother said that any baby not minding immediately would have been grabbed and eaten by the hawk.

Although this book has been through years of "good intentions" in the writing, Mother remains in my mind, even more clearly as I get older, and the things she told me become increasingly meaningful, her words still clear.

May 17, her birthday, is celebrated every year with chocolate cake and tea.

Dear Mother,

It's your birthday. You'd be 130 years old today, but maybe you're still 86 as when you departed, or perhaps still in your fifties, busy in the kitchen, like I saw you in a dream. Or maybe you are a young girl like my son saw you in his dream. I'll put the tea kettle on, get out the chocolate cake I made for you, and we'll share a simple celebration this afternoon as we always did. It's my childhood days I'm remembering you by today.

Mother, you worked so hard under harsh conditions at the mercy of adverse weather and a bleak economy during those Depression and Drought years to make a living off of prairie land where the wind continually blew, summer sun beat down, and winter blizzards were fierce; yet nothing ever fazed you from doing what needed to be done each day for the sake of animals or land. But whatever the situation, you always had time to repair your child's slight injuries or fix her hearty and healthy things to eat. You managed time to play with me as you did ranch work. You even dashed off a poem now and then; you were so talented. I sensed how glad you were to have a child, and all the hopes you had for me when I grew up. Even if life on the prairie was isolated, being miles away from other children and towns, you managed to give me a well-rounded education through your example, your knowledge, and the books you had from earlier years. Having been a professional nurse and a school teacher, you made sure I never lacked for learning, even if I learned spelling and arithmetic following you herding sheep, or reading and penmanship at the kitchen table by a kerosene lamp while you prepared the meal.

Yes, Mother, you always gave time to me, making me feel entirely secure. Now, looking back these decades, my appreciation grows each year. I suppose I never told you that I loved you; that was something taken for granted. In place of hugs, I playfully teased you by untying your apron strings or by running pell mell toward you only to stop within an inch of colliding.

I look back fondly on this early period of my life, when the prairie was my paradise and my companion and you were my role model and teacher. Thank you for being the devoted, perfect mother you were.

Barbara Ann

Part I

Opening the Trunk: Remembering the Early Years

Franz Willem Brunia

Allie Roorda Brunia

(l. to r.) William, Letitia, Sara, Bertha, and Henriette

Letitia:
She's Iron, Stone, and Beautiful Flowers

Years ago Aunt Henriette told my dad, as they were speaking about my mother, "She's iron, stone, and beautiful flowers." That does well to sum up Mother for she certainly had the strength of iron in her character. Her way of life portrayed a firm and strongly built foundation of stone, which nothing could topple. Beautiful flowers symbolize her sensitivity, and her talent for creating and appreciating beauty. Within her strength there was a gentleness, a soft kindness demonstrated by her deep love of nature in its myriad forms. She had respect and fondness for both tame and wild animals, and a love of plant life, especially the delicacy of the wild prairie flowers.

My mother, Letitia Brunia Black, was born the second daughter of Allie and Franz Brunia in Olivet, Iowa, on May 17, 1889. During her span of eighty-six years and three months she touched many lives and hearts. Letitia, a name meaning joy, came from a Frisian grandmother, Tietje, and is of pure Frisian lineage.

Her ancestors were proud, industrious people who in ancient times were pushed and crowded nearly into the sea by bigger, stronger warring tribes, but they could not be beaten down. With great determination and tenacity they clung to their culture and what was left of their land. They set to work to build dykes to hold the sea back, then laboriously brought up soil from the ocean bottom when the tide was out to make new land for their farms, cattle, sheep, and horses. Every acre they created was indeed prized. Such qualities of perseverance and stubbornness are Frisian characteristics. I would say that Mother rang true to the traits of her heritage, including her care and devotion to the earth. Her conservation practices demonstrated how precious soil was to her, which should not be eroded or blown away with the winds.

In 1847, her grandmother, Baukje (Bertha) Abraham Buwalda, immigrated to America with her parents and brothers and sisters from Friesland, one of the northernmost provinces of the Netherlands, as part of the Scholte Colony. This was a group of about 800 colonists seeking freedom from religious persecution in Holland for their views on the necessity of the complete separation of Church and State. After crossing the ocean in four ships and months of overland travel on foot or by wagon, they finally came to the Des Moines River valley in southeastern Iowa, where they settled in a large fertile area between Des Moines and the Skunk River. They named their new home Pella, a name taken from a Biblical city of refuge. At that time Pella was little more than a few log cabins in the middle of a vast prairie of endless grass.[1]

My mother's grandfather, Izaak Epke Roorda, came from Friesland to settle in the same colony in the early 1850s. All the Friesian immigrant heads of family were expected to have a useful occupation, such as rope making, cabinet building, blacksmithing, and farming. The families usually were large. Izaak and Baukje married in 1857 and had nine children, the second eldest being Allie, my mother's mother.

Allie Roorda grew up learning the skills expected of a pioneer woman: cooking, sewing, knitting and crocheting, soap making, cheese making, gardening, and preserving fruits and vegetables, among many other practical talents. She also had a deep knowledge of herbs and their uses for healing. In 1884, a young Hollander, Franz Willem Brunia, who had emigrated from Friesland the year before, came to work on her grandfather's farm, where he met Allie. They were married the following year, on March 4, 1885. They purchased a farm about a mile from her parents, where their five children were born: Bertha Ann, Letitia, Henriette, William Frans, and Sara Alice. [2]

My mother (Letitia) and her three sisters and one brother grew up together with thirty-one of their cousins and many uncles and aunts, who all lived nearby. Frisian was the language spoken in the home. The children also had to learn Dutch, and by school-age English was acquired. Mother used to get Frisian and Dutch newspapers, reading to keep her native languages active. When she and one of her sisters were together, they only spoke in either Frisian or Dutch (the latter language they called "Holland"). I always wondered what they talked about—most likely things they didn't want me to hear.

Mother's Trunk

Mother had a trunk she kept on the closet floor. It was the trunk she brought from Iowa with her most necessary and treasured belongings when she came to Colorado in 1916 to homestead. I was always curious what was inside. On the rare occasions the trunk was opened, I was always underfoot, full of questions about the things I saw. This never ending excitement was my adventure into Mother's early life. The trunk always had such a special smell. I don't know that I can describe it, because it wasn't only of moth balls, or cedar, or pepper to keep away the moths, though I'm sure she had pepper in all the wool things; it just had its own individual trunk-scent. It had a full bottom compartment, and a shallower drawer that was the top layer with several divisions. All was neatly organized, and cloth items carefully folded, enabling the trunk to hold more things. She had special letters, certificates, and valuable land papers and maps in some top compartments. In Mother's trunk were a few photos, some with ocher shade and a couple of the old tin types, of uncles, aunts, brothers, and sisters, her father sitting against the trunk of a tree, and

Opening Mother's trunk.

her mother knitting in a rocker. This typical homesteader's trunk served the purpose a safe would nowadays, for even if not fire-proof it was at least mouse-proof.

One day Mother had the trunk open to get something out as I peered in. Among fans and tatting, a small Frisian china doll caught my attention. As I picked it up to hold close, I asked, "Tell me about when you were a little girl, Mother?" I had asked Mother about her girlhood before, but I was always eager to hear again, and maybe she'd tell something new.

Mother grew up with a sister two years older, Bertha, a sister two years younger, Henriette, and a brother, Will, four years younger. Later there was another sister, Sara, ten years younger. Each week everything in their home was thoroughly cleaned, and Mother, starting at four years of age, had the chore of dusting. She would be tense, fearful some treasured heirloom brought from the old country would slip through her young fingers. Frisians have an affinity for cleanliness, and the story is told that when the first wave of Pella immigrants boarded the four ships bound for America, they immediately cleaned each ship from top to bottom.

When Mother's chores were done, she could go outside, where she played much of the time by herself, humming softly like a contented kitten. A special memory went back to the huge, old nut trees that grew in wayside corners and the flowers, berries, and nuts she gathered in nearby woods. Her favorite memories had to do with apples—Sheep Nose, Ben Davis, Snow, and Maiden Blush varieties. She often gathered and buried caches of apples down under the fallen leaves, coming back later to find them mellow and delicious. Love of apples remained with Mother all her life. Her father raised fine hogs as well as the usual assortment of farmyard animals and grew fields of hay and grains. He always rotated a field of cow peas, which was planted to enrich the soil like alfalfa. Mother said she thoroughly enjoyed helping her father harvest corn in the fall by tossing the ears into his horse-drawn wagon. Her mother kept a large garden, made butter, cheese and soap, canned and dried fruits and vegetables, and cured the meat. By fall their cellar was well stocked with hundreds of jars of fruit and preserves, dried foods, and winter vegetables, such as potatoes, carrots, celery, turnips, and cabbages. Mother was surrounded from childhood by, and soon engaged in, the useful home and farm arts that were in later years to become an integral part of her own life.

I found a little book, likely a reader, and next to it was a roll of fairly large, cream colored art paper tied with a narrow, light blue ribbon. Curious, I untied the ribbon to find they were paintings of apple trees in bloom, roses, even Dutch scenes. "Oh Mother, did you paint these in school? They're so pretty! What was it like to go to school when you were young?"

Schooling started at four years of age for Mother, since her mother required her to accompany her older sister Bertha as they walked a couple miles to the country school. They learned reading from the McGuffey books in use at the time. This period was not a happy memory. Mother had to share the same desk with her sister, who had a nervous disorder and often pinched Letitia unmercifully hard. Mother was timid and never dared to even utter a peep. Bertha and my mother were about the same size, so their mother dressed them alike, and they shared the same bed at home.

High school was attended in town. It was during these years that Mother's artistic abilities flourished. She made excellent botanical drawings of all the plants in their area and kept an herbarium of dried pressed plants and flowers, which still remains intact over one hundred years later. Mother also had an ear for music and played the violin with a sweet touch. She wrote the high school class song to the tune of "Redwing", an old time favorite.

Carefully wrapped in pillow cases with tatting edging were Mother's china paintings. Embroidered napkins protected smaller things. "Mother, it would be so much fun to have a tea party with these," I said, as I held up two little tea cups. "Not today, maybe sometime we will."

In 1906, during her junior year in high school, Mother did considerable painting and illustrations; she also painted china. She had painted a tea pot and six cups to accompany it with a stork-like bird that had a long neck, fluffy white feathers, and a black tail. We never used the set, but nonetheless nowadays it is proudly displayed. Other fine china paintings on vases and plates are also her works of art that I show with pride, saying, "Mother painted these." Two vases, each a foot high, grace the top shelf in my bedroom. One is a collage of old cigar wrappers, shellacked to keep the wrappers shinny and safe, and the other is elegantly painted with bluebirds. Many of her lovely

Teapot and cups painted by Letitia.

watercolors, which I found rolled in the trunk, are now framed and hang in our home.

Sadly, because Bertha was jealous of her accomplishments, their mother soon forbade my mother from painting and playing music in order to keep the peace, so these talents became latent within her. Bertha was prone to having tantrums and crying spells, which her sisters were expected to humor and go along with. To others Bertha was charming and a pleasant conversationalist, but to her family she often showed a different personality.

Mother had other things to do and was ready to close the trunk just as I picked up a red valentine with cupids, arrows and hearts, all in a one piece cutout from a red sheet of paper. "How did you ever make this, Mother?" I held it up, thinking it the most beautiful valentine ever.

The summer after she graduated from high school in 1908, Mother took "normal training", as they called studying for teaching school back then. She passed the required examinations for a teacher's certificate with excellent scores that same summer and was hired to teach at a one-room rural school near Prairie City. The following summer, her sister Henriette also obtained a teaching credential, and both taught in adjoining rural schools near Prairie City. It was a pleasant year for the two sisters.

For the school year of 1910–1911, the two sisters transferred to schools near Olivet, where they boarded with their aunt and uncle, Mr. and Mrs. J. A. Herny. Some days Letitia had as many as 64 pupils ranging from first to eighth grade, with some boys bigger than her and nearly as old, having taken longer to get through school due to being out to work. As well as creating lessons for all grades, Mother also devised many arts, crafts, and games suited to her students interests and abilities. The children at this new school were a mixture of rural farm children and those of the town's coal miners. Mother would often be faced with the dilemma of keeping peace and harmony between these two culturally different groups, but she tactfully managed to build many lasting friendships among her pupils and their families. The children were always proud to have their teacher come visit and have supper in their home. One of her sweetest memories was about a poor miner's little daughter, who used to come and lay her cheek against a soft pink, chiffon blouse Mother often wore especially to please this child. The older boys were respectful and helpful to their young teacher, often assisting with the heating stove and other manly chores. At a school box supper, Mother made her meal in a pumpkin, surprising the man who won the teacher's "box".

Months went by before Mother again opened her trunk for something in the top compartment. In one of the divisions were photos. "Who are these people, Mother?", I asked, as I held up a group photo. Mother explained that they were nurses. "These were my sister-nurses when we all graduated together in our caps and white uniforms." She pointed each out by name. "This one is Flora Smith and that's Barbara Padgett, whom I named you after. Barbara was my most loyal friend, so you are her namesake in admiration of her."

Besides the photos—rolled in a soft, cream colored woolen cloth—were fastened the pins and other nursing "jewelry" Mother had worn on her uniforms. In one of the compartments was a not-too-large black leather case that opened out with pockets for surgical things, such as her scalpel, and leather loops to place thermometers and special curved sharp scissors. Mother unrolled some large papers to show me her diplomas and state board nursing certificates.

One of the Herny's daughters was nearly the same age as Mother, and both decided they wanted to become nurses. So that fall (of 1911)

Letitia, the nurse

the two girls entered a nurses training program at Mahaska County Hospital in Oskaloosa, Iowa. Although her friend had to drop out because her mother became ill, Letitia finished her training and became a fully qualified RN. Mother excelled in this work, always taking the best care she could of all her patients. After working as a nurse for several years, she was asked by Dr. Williams to become the superintendent of a new private hospital in the same city. She was in her mid-twenties at this time. All of the nurses there were in training, so among her duties, she was responsible for conducting classes for the student nurses. From her own experience and nursing textbooks, she did her best to train them well. Wanting a spotless hospital, she frequently checked the premises. Whenever she found the least trace of dust, she would leave her initial "L". Upon seeing the sign, the young nurses were quick to clean especially well, being always eager to please their gracious superintendent.

My mother made many life-long friends during her nursing years. During this time, her two sisters, Bertha and Henriette, also decided to take nurses training. They graduated from Mercy Hospital training school in Des Moines in 1915. In those years nurses worked long hours, often twelve or more hours a day, especially when many people were ill, as in epidemics. Due to her many and weighty responsibilities, my mother often felt tired and very weary.

Homestead Decision

The trunk held all of Mother's land papers and land maps with the section numbers. Mother promised to tell me about them some day when I was older. I didn't bother further; they didn't appear all that exciting to me anyway.

It was also during 1915 that mother and her two sisters learned of homestead land available in Colorado. One day some friends came back from Colorado, where they had taken up homesteading a claim, and expounded on the wonders of their new life. This appealed to the sisters' sense of adventure as well as their desire for a more restful lifestyle. Bertha and Henriette traveled that summer to Fowler, Colorado, where they stayed on the ranch of Harry and Nella Fassett

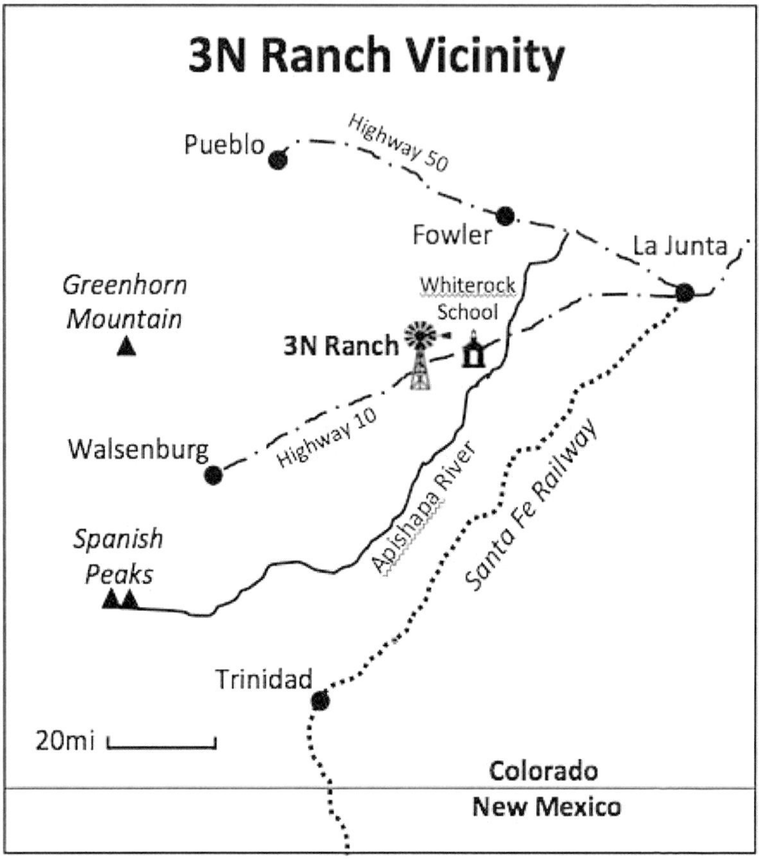

and were introduced to Colorado style ranch life, which they found fascinating. There was still homesteading land available on possible oil fields, including mineral rights. The two sisters, persuaded this was an ideal opportunity, wrote to Letitia and asked if she would be interested in the three of them each acquiring a section of homestead land.

As for modern day women's lib, Mother was one of its forerunners, by example if not by speech and attending rallies. Their father, Franz, was set in the ways of Frisian tradition by which he had raised his daughters. He became terribly angry once when Mother donned bloomers to ride with her brother, Will, on his motorcycle. He was quite a daredevil as a boy, and Mother was sure he got the motorcycle up to sixty miles an hour (very fast for those years!).

Franz was adamantly opposed to his daughters' desire to go homesteading on their own. It was unheard of and an embarrassment to him. He yelled, "No single young women should ever do such

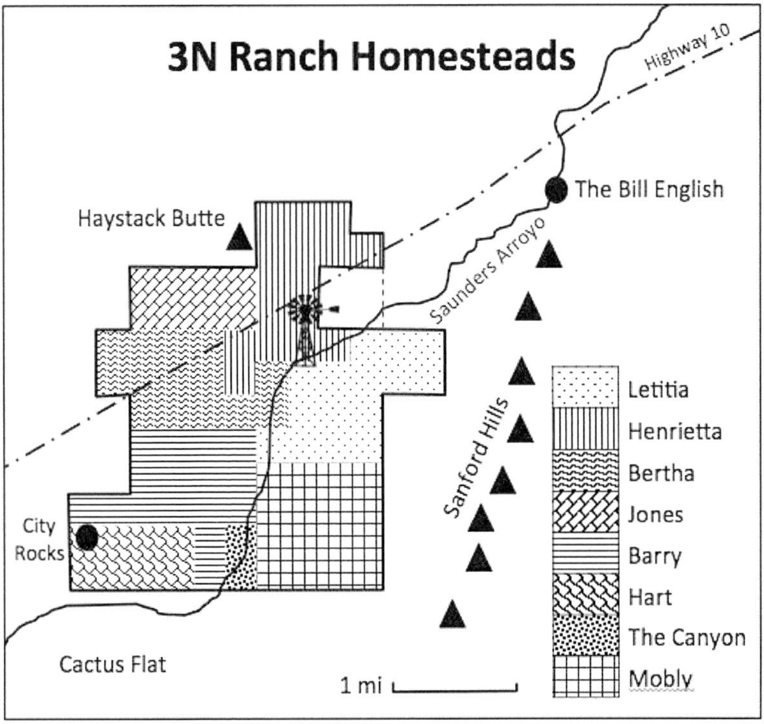

things!" Certainly, his daughters would not break with the expected roles of womanhood. But Mother had been on her own since teaching school at nineteen, had been a nurse and superintendent in charge of other nurses, and she would be responsible for her own future. Her sisters had lived and worked independently as well. Didn't we say stubbornness was a Frisian trait? Undeterred by their father's wrath and wishes, they went anyway, leaving Iowa by train, each taking what possessions fit in a trunk, destined for Pueblo, Colorado, and the unknown, to a still wild spot of prairie land in Colorado. Each sister would get three hundred and twenty acres of ground, which if she made the improvements according to the law—building a cabin, digging a well for water, and plowing forty of the acres for farming—would be her very own through the Homestead Act of 1909.

After getting off the train at Pueblo, Colorado, Mother and her two sisters went to the county land office to check on homestead requirements. They were able to have a "prove up" hearing for three 320 acre sections of land cornering each other. After the paperwork was completed, their new friends, the Fassetts, met them, trunks were loaded in the back of a horse-drawn wagon, and they were on their way to a new and vastly different stage in their lives. The Fassetts helped them locate an area of uninhabited Colorado prairie the sisters thought desirable to make the claims. The land they chose centered on an arroyo, where underground water could be found. It was an ideal place to have the three homestead sections adjoin and on which to build their cabins fairly close to each other. The arroyo, known as Saunders Arroyo, was 45 miles southeast of Pueblo. Surveyors then surveyed the land to stake the homesteads, and the claims were filed in the Pueblo county seat. During the surveying, the sisters stayed with the Fassetts and began to make plans for moving onto the homesteads the next spring.

Once when Mother needed something from the trunk in a hurry, and of course I was always right there in the way, I saw a black leather photograph album and wanted to look at it. "I'm in a hurry now, but we'll leave it out, and tomorrow night I'll show you and Dad all the

pictures and tell you about this place and some of the things that happened when my sisters and I first came." (Mother always called my dad "Dad." I never heard her call him by name.)

3N Ranch: Three Nurses Build a Homestead

In the early spring of 1916, Letitia and Bertha left Iowa and nursing temporarily behind to come to Pueblo, where they stayed at the YWCA. Now began the many things necessary to be completed: Their cabins built, forty acres to be plowed and planted with a crop that depended only on rainfall, termed a 'dry land crop', a well to be dug, and a cellar for food storage as well as protection should a cyclonic whirlwind strike their cabins (a fate of many other prairie shacks). A cistern, shaped like a huge underground jug, was another vital homestead need to store drinking water and keep it cool during summers as well as protected from freezing during winters. The circular pit, about eight feet wide and ten feet deep, was dug by hand close to Mother's front door. A fellow homesteader was hired to construct a wooden form to pour in cement to make four-inch thick walls. The walls curved sharply at the top into a three foot diameter, cylindrical opening, eight inches of which stuck above ground for the wooden lid.

They first bought a Model T Ford, learned how to drive it, and with its manual to fix problems. It had to be cranked to start, could "kick back", and had other difficulties these previously non-mechanical

ladies had to overcome. Having known only horses, this thing didn't understand whoa or giddy up! To avoid the embarrassment of being seen, they practiced learning to drive on the prairie.

They hired homestead men of the vicinity, glad to have some paying work, to build their cabins, to put up the windmill tower for the well, and to do the required plowing. The first house built was on Letitia's land. It was a simple structure with two rooms, twelve-inch wide plank floors, tar papered walls, and a bow-shaped, sod-covered roof. Sod was available at no cost, was excellent insulation and fire resistant; it helped keep the place cooler in summer, could withstand the prairie winds and thunderstorms, and needed no skill to apply. One of the rooms served as a kitchen and other as a bedroom. Linoleum was later put down in the kitchen over tar paper and a coating of bran to level out any rough spots or bumps in the plank floor. The bedroom

Letitia's two-room sod-roofed cabin, as it appeared 100 years later.

had twelve-inch boards, laid close together so the cracks between were narrow.

After Henriette arrived, the sisters moved to the ranch and lived in the first cabin while the other two cabins were being constructed. Henriette's cabin was a small one-room "shack", while Bertha's was a nicer frame house consisting of a large living room, two closets, and a wash room, with a garage for the Ford at the end. The living room had a large cobblestone fireplace with benches beside it, book shelves, and a large table, all made by the carpenter. Small pane windows up high on the north and east sides allowed in extra light. All the homestead windows in the area had small panes, as they were much easier to replace after the damage caused by a summer hail storm. Hand braided woolen rugs were placed on the wooden floor and Letitia's water colors on the walls. Her cabinet phonograph with a collection of classic records, books, magazines, and knickknacks created a warm and inviting atmosphere in the room. This house was later moved and joined to the one on my mother's land to make the present ranch house.

Bertha's house

The windmill was erected, picket corrals were made with accompanying shed, and fencing was put in to be ready for the purchase of livestock. Soon after their arrival, the homestead law was modified once again, under the Stock-Raising Homestead Act of 1916, which allowed them to double the size of their original allotments from 320 acres to 640

acres each. Some cattle and one riding horse were purchased from the Fassetts.

Once the ground was plowed, what to plant? Henriette had seen wild gourds growing near her plowed spot and decided on pumpkins, something they could eat, trade to neighbors, and nutritious feed for their cattle. Mother planted milo maize, or sorghum, for winter cattle feed. Once this obligation was met, there was no future planting, and the ground was left to re-sod against the winds that stirred the soil to sail away. Even a hundred years later these areas have not completely healed and continue to grow tumbleweeds.

Their savings were already dwindling, not sufficient to enable hiring all the labor needed for the many things yet to be completed for "proving up". Their once soft hands, blistering at first, soon became well callused. When the well they were digging by hand (with help from Harry Fassett and his brother-in-law) got too deep to toss the dirt out, one sister sat on the bank, reading between bucketfuls, to

The ranch as it looked in 1916.

bail up the pails of soil while the other dug in the bottom, exchanging places now and then. The Model T came in handy to haul many loads of heavy sandstone slabs from the canyon, a mile away, for curbing the well. More loads of sandstone slabs were brought in to curb the walls of the cellar they excavated next. The Model T enabled them to scout around the country while looking for saddle horses and cows to buy, so the ranch would in time begin to make a living.

Once cows and horses were purchased, it was required to obtain a brand from the Colorado State Brand Office. It needed to be renewed every five years. In 1917 there were already some 42,000 Colorado

Whiteface cow with the 3N brand.

stock brands; so it paid off to never let one's brand lapse. It was by good luck that a reversed 3 N brand was assigned to the Brunia sisters, since they were commonly referred to as the three nurses, and their homesteads had collectively become known as the 3N Ranch.

Letitia named each cow.

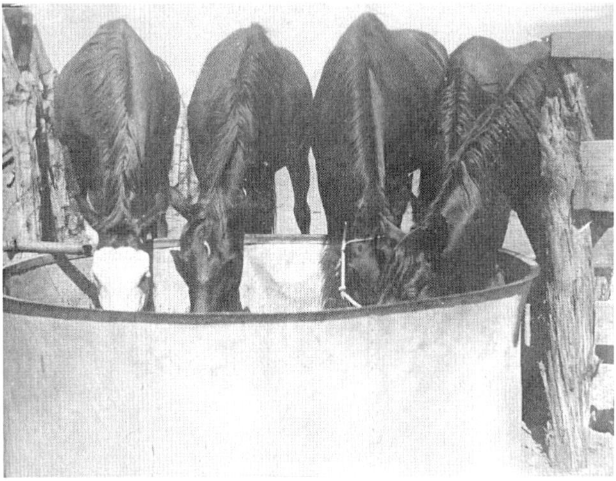

Three Nurses' horses having a drink.

Mother and Babbie

During World War I, Bertha and Henriette enlisted as Red Cross nurses. Henriette was assigned to Fort Riley, Kansas, and Bertha to an Eastern army post. Letitia was left in charge of the ranch, because she was considered by her sisters "by far the most capable" of taking care of it. Later Bertha went overseas, in place of Henriette, as an army nurse. After her service at Fort Riley, Henriette returned to the ranch in the summer of 1919. She reported in a letter to the author written in 1974: "Letitia had the ranch under excellent control. The stock had increased in number, and all were still being named—many with fancy names such as Harmony, Melody, Louise, and Angela. Our mare had foaled a colt, an unusual light sorrel with a whole white face, who was named Iowa."

The youngest Brunia sister, Sara, was visiting the ranch at the time. She had been ordered by her doctor to have a rest, and what better place than a ranch in Colorado. Sara had also become a registered nurse. From the very beginning the 3N Ranch was a popular vacation destination for friends and relatives from Iowa. Their guests the first summer in 1916 included Clara Jackson from Pella and Charlotte Keating, a young girl from Oskaloosa. Clara's visit included a surprising romance between her and the Brunia sisters' neighbor, Carter Hagens, who came over to entertain the visitors with his tales of Western life.

To bring in money for the 3N Ranch, the sisters took turns leaving for outside employment. In the spring of 1920, Letitia went to Iowa to do nursing, and Henriette stayed behind to tend the ranch. At this time, a new neighbor came by in a pickup and introduced himself as Bill Barry from Smith Center, Kansas. His homestead section adjoined Letitia's and Bertha's to the south and west. He was eager to be of help, and soon made himself almost indispensable around the ranch by keeping the windmill oiled and in repair, fixing water gaps after rains flooded the arroyos, chopping wood, etc. In exchange, Henriette was pleased to bake him bread and often invited him to stay for meals.

Bertha now only spent short amounts of time at the ranch. Because she and Henriette were veterans, the Homestead law required them to spend a limited number of months on a claim to prove it. My mother, on the other hand, was required to spend at least three years on her claim to prove it. Henriette and Letitia wanted to both be able to stay on the ranch while still making an income, so she and Letitia discussed a plan to make this possible. Henriette obtained a contract from the district school board to teach at the one-room White Rock school starting in September 1921. She rode the nine miles to the school by horseback each day, alternating riding on Babbie, Iowa, and Sam. Letitia, now back from Iowa, took over the ranch duties. She was later elected to the school board, so both sisters became more and more involved in the community.

In early 1922, Bertha asked to no longer be a partner in the stock raising. The cattle had to be divided, and Bertha's cattle were sold. It would have been difficult to divide them if my mother, according to Henriette, had not known every one of the animals by name. Bertha married Louis Unger on April 3, 1922, and moved with him to Wheatland, Wyoming, where they became the parents of three sons: Clayton, Ellis, and George.

In the fall of 1922, both Mother and Henriette left the ranch to work, leaving it in the hands of Mr. Thrap. Letitia nursed at Corwin Hospital in Pueblo, and Henriette nursed at Skiff Memorial Hospital in Newton, Iowa, where Sara was also a nurse. By the next spring the two sisters returned to the ranch. It was not long after that that Henriette became engaged to William (Bill) Barry, and they were married in Fowler on October 15, 1923. Once again a division of the livestock was made. Henriette and William went to live on a farm near Stratton, Colorado, where their two daughters, Alice Elizabeth and Letitia Anne were born, the latter daughter being named after my mother. Through the years the Barry family made frequent visits to the 3N Ranch.

After her sisters married, Mother remained at the homestead and eventually bought the land that had belonged to her sisters so that the three allotments remained intact as a single property. She too married in 1924. That story follows in the next chapter.

Mother and Buddy

There were some very severe blizzards during the 1920s. One night during such a tempest there was a knock on Letitia's cabin door. Even the hair on the back of her dog, Buddy, stood on end. She could not imagine how anyone could be out in weather like that: below zero with howling cyclonic wind whipping the snow around. She opened the door to the stranger, who turned out to be her horse, Iowa, who spent the rest of the night in the kitchen. From that night onward Mother always left a lamp lit, sitting close to the window, as a beacon to anyone lost in a storm. In winter, when darkness falls early, and especially in a snow storm, a rider could easily lose his way.

Mother got lost in a snow storm once. It was evening and she had most of the sheep and cattle in the corral. It was stormy and she decided she'd better look quickly for the rest that weren't in. Darkness overtook her and soon she was hopelessly lost. It was snowing and there were deep drifts. She began to consider how she'd spend the night. She felt she'd keep most comfortable if she burrowed into a snowdrift and stayed until morning. Just then she heard sheep bells and thought perhaps it was someone moving sheep, but that was hardly possible! Walking a bit farther in the direction of the sound of bells, she stepped over the edge of the arroyo and fell into a snow bank. She discovered it was the arroyo just below the corrals, the bells were from their own sheep, and she was safely home.

After another heavy snow a cow was missing. Mother had searched everywhere not finding any trace of her. Walking along a draw she all at once had a strange feeling, as if eyes were watching her. Looking closely into a deep snow drift, she saw two eyes peering out at her. The cow was completely buried except for part of her white face. Evidently she had sought shelter behind a low bank and lain down. Heavy winds during the night blowing the snow around had created a deep, heavy drift completely over the cow as she slept. Zero temperatures had frozen the drift so solid the cow was unable to get up.

Mother told of other incidents about her early years on the homestead. Sadly, I remember very few. If only I would have been more attentive and asked more questions about the ranch's first fifteen years prior to when my own memories there begin. The two following stories were related to me by Aunt Henriette.

Mother on Iowa and Henriette on Babbie.

Once Aunt Henriette riding Babbie and Mother on Iowa had a long and hot ride. When they came to some water holes, they decided to let the horses drink. But Iowa, with Mother on her, decided instead to lie down and roll in the water. Mother clambered out of the water, got Iowa to her feet and exclaimed, "Why, Iowa, you naughty baby!"

Mother had a gold watch, one she used when a nurse. One time she and a sister were riding on the open range, when she discovered she had lost her wrist watch. They searched carefully trying to retrace their steps, but to no avail. Mother always had a feeling it might show up sometime. A year later Claude Hart found the watch and brought it to Mother, thinking it might be hers. In spite of dust storms, rains, snow storms, freezing and thawing, winds and more rains, all the watch needed was winding to resume its ticking! In the vastness of the prairie that certainly was like "finding a needle in a haystack".

Ranchers know the meaning of friendship, the neighborliness of helping out, or just dropping in to say hello and inquire about everyone in the family. In the homestead years, when people lived miles apart, telephone lines didn't exist, and transportation was mostly on a saddle horse, or team and wagon, neighbors felt an even stronger bond to each other. Mother was no exception. She would stoke up her wood cook stove and have a chocolate cake or pie baking in minutes as they talked. No company ever left without sitting to a hot cup of tea or coffee and baking still hot from her oven. Sharing troubles and trials was a way of shedding stress, and if help was needed, naturally it was offered with no thought or hesitation. It was customary for a male visitor to chop several armloads of wood in appreciation for a meal.

The "three nurses" have passed on, and most of us cousins as well. Our parents would tell us about their early homestead beginnings and unusual happenings, but none of us wrote things down, trusting to memory. Over time, with moves and downsizing, few of our parents' homestead letters remain as a source of history. It is thanks to Aunt Henriette, who wrote down many details about the Brunia sisters, both from their childhood and early adult years in Iowa and from the homesteading period in Colorado, and sent them to me before she passed away, that the preceding history has been preserved.

The 3N ranch endured, while many of the neighboring homesteads were eventually abandoned because their owners could not make a living off them. Not every aspiring homesteader was as fortunate as my mother and her sisters were to have a reliable source of water on their land. Others left for other reasons.

Mother was most protective of this land she loved. Her conservation efforts were unusual in her time. She helped to prevent erosion by not permitting her cattle to overgraze; she filled small gullies with tumble weeds or ranch discards to prevent them from becoming mini arroyos. She brought pickup loads of manure from the barn to spread on the less fertile areas above the corrals; she dug contour ditches on the steeper slopes to support the growth of indigenous plants. Her preservation and improvement of the land was not only for herself, but she wanted to benefit future generations. Her motto was: always leave something better than you found it. The following poem, written by Mother the day after she got back to the ranch after being absent for several months in the hospital with a broken hip, illustrates her love of the land of which the 3N Ranch became an integral part.

Three N Ranch
by Letitia Black

This is where I like to stay,
Stars at night, sun by day.
Breezes that are sheer delight,
Winds that sometimes are a fright!
Grasses vivid green in spring,
Hot July another thing.
Lupines, azure as the sky,
Then tumbleweeds go sailing by.
Cacti blooms that thrill the eye,
Spines that make you pass them by!
Clear, clean air, our lungs expand.
Next they're choked with dust and sand.
Space that gives the eye a fill,
Blizzards cut that view to nil.
Baby calves that warm my heart,
Tears I shed when we must part.
Horses are my heart's own love,
But not if tumbled from above!
Food never tasted half so good,
But first one has to chop that wood!
Well, anyway, life can be most fine
Though times be grim, they're oft sublime.
The elements may batter and sear.
But this is my home, so very dear.

Mother and Dad Meet

Mother had not put the photograph album back in the trunk. Coming up next were pages with pictures with my dad. The next night we again sat at the table, the kerosene lamp in the center, to look at photos. One earlier time when the trunk had been opened, I had noticed Mother's and Dad's wedding certificate. It was dated August 24, 1924.

My dad, William Marion Black, was one of the West's colorful pioneers in the years before my birth. The early years of his life in Idaho are a chapter in the building of the West. The circumstances of his birth, September 14, 1866, were unusual. His parents, Charles Morgan and Annis Matilda Black, were members of an oxen train en route from Chillicothe, Missouri, to Boise, Idaho, when the wagon train stopped for my dad's birth. The wagons were formed into a circle as a defense against a possible Indian attack. The Indians (probably Shoshone or Paiute) did come and made it known that the wagons had to move on, taking an uncle hostage until the wagons moved. They allowed only my dad's family to remain in that vicinity. His parents stayed, built a log cabin, and opened what became known as Black's Station (or Black's Creek Station), from which they operated a stage coach between Boise and Silver City for sixteen years. Black's Station, about sixteen miles east of Boise, was the usual breakfast stop for travelers heading east. During this time, with no other white families around, Dad often played with the Indian children and rode with them bareback, though he was not lacking for companionship, being the fourth of nine children born to Charles and Annis Black.[3]

Charles Morgan Black and Annis Matilda Black

Courtesy of BJMEANS, Ancestry.com

In 1882, his parents moved to Bellevue in the Wood River valley, where they farmed, ran cattle, had a dairy, and raised wild horses and hogs, furnishing meat and dairy to various mining towns. Sometime after his mother died in 1894 on their ranch near Bellevue, his father remarried an Irish woman named Delia Burgan. They lived on the ranch for several years before moving to town. Dad then purchased his father's home and ranch, and also bought the neighboring Bellinger ranch. Edward and Martha Bellinger, who had come to Wood River Valley in a wagon train from Oregon, had arranged with the Blacks for their daughter, Etta Bellinger, to marry Dad in 1893. Etta was actually in love with Andy Liggitt, but her parents had considered Andy Liggitt too poor and a harum-scarum. Dad, on the other hand, was a well-to-do rancher. After their marriage, Etta was enamored with high society. Their two girls were sent to finishing schools, Mary to New York and Violet to Europe to learn painting. Their son Charlie was sent to Catholic boarding schools, but he constantly ran away. Jeff was a daddy's boy, who loved the ranch and wanted to stay on it. After their children were grown, Dad, knowing of Etta's love for Andy, arranged for a friendly divorce. He bought Etta a house in Boise, and Etta eventually married Andy.

Violet remained with Dad on the Idaho ranch, helping with the bookkeeping and packing butter, eventually becoming the manager of the family business. She married a doctor, Fred Griffith, six months later, whom Dad liked. Charlie married a girl named Margaret, and they lived on and ran the Roughbottom section of the ranch, while Dad and Jeff ran the lower place. Mary was not interested in the ranch. She later married an influential Berkeley lawyer, Leo Wilson, and moved to California. There was some jealousy and hurt feelings between the children. Charlie wanted to be with his dad like Jeff and was jealous that Dad got along with and talked to Jeff the most. Violet was hurt that Dad seemed to favor and spoil Mary, his youngest daughter, while Violet was doing the hard work of managing the ranch. Once he agreed to buy Mary a $500 fur coat, knowing Violet was going to get married in two weeks. Another time Violet threw a chair at Dad because he wouldn't make Mary help hang up the sheets, and he laughed when she just asked if her breakfast was ready yet.

The family business was making money. According to James Hawley's History of Idaho (published in 1920), William Black "prospered in his undertakings…and kept adding to his place from time to time until he now has five thousand acres of fine ranch land and is one of the leading and progressive ranchmen in this part of the state," as "manifest in the excellent appearance of his ranch property and the splendid improvements he has put thereon."[4] According to Violet, the value of Dad's property at that time was $75,000, ($1,000,000 in 2018 prices) not including the price of his 5,000 head of cattle. But then the Depression of 1920–21 hit. It was a period of extreme deflation following World War I. Wholesale prices dropped by 36% and surviving businesses saw a 75% decrease in profits.[5]

Margaret wrote that Dad had "expanded in land too fast and everything was heavily mortgaged." In addition, that winter was long and harsh, and the price of hay went from two dollars to fifty dollars a ton. The family had to trail their cattle sixty miles to feed them. Unable to pay for his debts, Dad was forced to go bankrupt, and they lost everything, only managing to salvage some household furniture, six horses, and a few cattle.[6] I remember him being upset with Republicans (he was himself a Democrat), placing some of the blame for his loss of the Idaho ranch on Republican policies.

In 1922, Dad moved to the Mustang area south of Fowler, Colorado, where Cliff Stauder offered him the position of managing one of his sheep ranches. Both of Dad's sons went with him. Jeff, who was still single, lived with Dad and helped him on the sheep ranch. Charlie, his wife Margaret, and their daughter Betty, along with their famed race horse, Darby, moved to Rocky Ford (18 miles southeast of Fowler). Violet and her husband Fred moved to Reno, Nevada, where he set up a successful medical practice. Charlie's family decided they couldn't make a new start in Colorado. They soon relocated to Los Angeles, where he worked in construction for almost nothing until a lucrative position opened up for him in the oil fields. Jeff remained in Colorado, working with his dad on the sheep ranch. It wasn't long after their arrival that Dad heard of the three homesteader nurse sisters, some twenty miles to the south, and went to meet them. Two of the sisters, Bertha and Henriette, had already married and left. By August 1924 he convinced Letitia Brunia, my mother, to marry him. He was 58 years old and well built. Mother was 35 and had always wanted children, and she knew her child bearing years were waning. She considered Dad, although much older than her, more suitable than the rough homesteaders who had proposed to her earlier. Except for Jeff, his second oldest child, Dad's children from his first marriage (my half-brothers and half-sisters) were by now married and raising their own children. Dad's grandchildren were older than me by the time I was born.

Mother and Dad

Mother, Dad, and I continued looking at the photo album, but there were not many pictures of them at that time, so I asked Mother and Dad to tell about when they were first married.

They were married by a justice of the peace without a big ceremony, and later the neighbors

Dad herding sheep at La Veta below the Spanish Peaks.

Mother in the sheep wagon.

surprised them with a shivaree, a form of good-natured celebratory hazing commonly enacted with newlyweds in the Midwest. Dad lived with Mother on her homestead after their marriage while still managing the sheep ranch where Jeff stayed. Mother told me that they later ran the sheep by La Veta, at the foothills of the Rocky Mountains below the Spanish Peaks, and lived for a summer in a sheep wagon. Her youngest sister, Sara, also a nurse, came out from Iowa to visit at that time. After meeting Dad's son, Jeff, they fell in love and got married not long afterward. Their oldest son, Jeffery Jr., was born two months ahead of me on October 16, 1928.

Heading for California to "Make Their Stake"

Since Mother did not make much income from her homestead, nor was Dad's ranch management very profitable, they decided to go to California for a few years to "make their stake" as they termed it, and then come back to live on Mother's homestead ranch.

I pointed to some more photos, and Mother said they were of when they left for California in late 1927, taking only what would fit in their Ford car and cutting down the back seat for sleeping on the trip. Jeff and Sara joined Mother and Dad in relocating to California, and packed their belongings in their own car for the trip. In one picture, they are sitting on the running board studying a road map. Bill Henderson, the

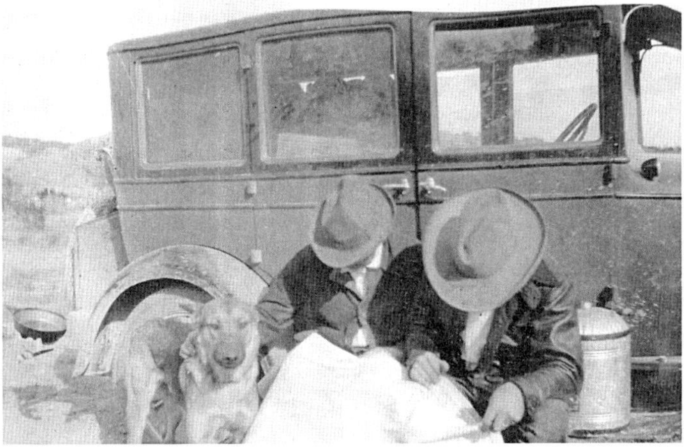

Jeff and Dad looking at road maps.

Sanford Ranch buckaroo who you'll meet in a later chapter, is pictured on horseback seeing them off. Next were pictures of their destination, the Gift Ranch, near Kneeland Prairie, California, (about 12 miles east of Eureka), where the two families lived and worked in 1928. It was a beautiful cattle ranch with several enormous barns surrounded by tall redwoods. There were also pictures of me and Jeffery as babies.

Bill Henderson on Midge

I was born on December 11, 1928 at Mrs. Bernier's Maternity Home in Eureka, where Mother told me we both nearly perished due to her long, difficult labor. Letitia, now nearly forty, dearly loved her first born baby. She would carry me out daily into the peaceful redwood groves to let me nurse when I wanted or nap in a cradle of redwood duff. Mother wrote that my baby eyes were "so expressive they fairly talked." During this time, Dad's oldest daughter, Violet, and her husband, Fred, came to visit us in Kneeland. At four months old, I started having digestive problems and was losing

Dad holding me.

weight. Mother was extremely grateful to Dr. Griffith for devising a formula that soon cured me.

Mother wrote in my baby book that Little Jeff got a cart we both loved to ride in and that I would sit behind Jeffery and got so excited I pulled his hair. I had learned a new word "iddy" and got upset when I was not understood. Finally, Mother and Aunt Sara figured out that I wanted a ride and was trying to say "Gitty-up". One of my treats was to sit in the cart and eat a graham cracker. My favorite animals were "cowies" and "kitties" but especially "horsies" and "doggies". Dad bonded with me through horses, loving to hold me up in the saddle on Goldie's back.

Living on this wonderful ranch was not to last, as the owner's son, who had moved to Montana, later changed his mind and decided

My first horse ride.

to come back, so my parents had to move. It was September 1929, I was not yet one year old, and Mother and Dad had not yet "made their stake", nor were they ready to return to Colorado. They decided to go to southern California, where Dad found an opening in the

Little Jeff next to Barbara in his cart.

Little Jeff and I eating our first ice cream cones.

stockyard business. Aunt Sara, Jeff, and Little Jeff stayed behind in Kneeland Prairie. For the journey, Mother and Dad equipped the car for camping. We slept in the car and ate meals outside. On the way, they visited Dad's son Charlie, "Auntie Peg", and Betty in Orange. Then they traveled to Oceanside, where we camped near the beach and I learned to crawl in the sand.

One day Dad took me to the beach to play in the sand. He thought I was the most active toddler he had ever known and needed an outlet for my energy. A couple walked by and asked him if I were his grandchild, as Dad was over sixty. When he said that the combined age of her parents was over one hundred years, they walked on without responding. It was not considered proper in those days for an older man to marry a much younger woman.

From Oceanside my folks took a trip into southeastern California's Imperial Valley, where they stayed for a short time, before heading west and north to reach Buena Park (not far from Orange), where they rented a house on Homewood Avenue. Dad got a job near here feeding cattle in the stockyards. (It may have been a dairy, as Buena Park was first known as a dairy center and a hub for both the Santa Fe and Southern Pacific railroads. The Los Angeles Union Stockyards in Vernon were about twenty miles north of Buena Park.) Mother's photo album didn't have any pictures of where Dad worked in southern California, and it never occurred to me to ask him where the stockyards were located, what it was like, or what he did there.

Jeff, Sara, and Little Jeff came down from Kneeland in December to visit us, and they joined us to celebrate Christmas at Charlie and Margaret's place. For my second birthday, Mother baked me an angel food cake. She wrote in my baby book that when she asked me if I would start my dinner with milk toast, I replied, "No, I'll start with birthday cake."

I was too young to remember more than two of my own experiences in Buena Park. The first one was trying to fly. We were at an elderly lady's home who we called Grandma Dillahoney. She had some chickens that flew across a little ditch in her yard. I wanted to fly over the ditch too. So I flapped my arms to fly like they did, but my wings didn't work, and my plump little body landed in the ditch instead of easily over it. I got slightly bruised and cried. I cried much more from disgust at not being able to fly than from the scrapes and scratched knee hurts. I told Mother something like I needed "fedders" (feathers). Mother and Grandma Dillahoney were kind and sympathetic, and didn't laugh at me, as I was very sincere and earnest, but I can imagine that they had a hard time restraining their laughter. Landing in the ditch didn't deter me from trying again.

When I was two, I'd often talk about "the man in the hole", who obviously had impressed me. Every time I told Mother, it gave her no end of puzzlement, for she knew it was something I'd seen and experienced, but from only that much description she couldn't figure it out. Yet I often talked about it, and Mother continued to be baffled. Not until later on did I describe the situation in enough detail that Mother, to her relief, finally figured out what I was talking about. It was a merry-go-round. It had horses that went up and down, and the man responsible to keep everything functioning properly was standing in the center below the platform. So he was "the man in the hole"! The man had been important to me, as I believed he made the horses go up and down for the ride.

The Great Depression

The Great Depression began on October 29, 1929, Black Tuesday, with a crash of the stock market. My parents were likely unaware of how bad the economy would become, as business continued, although many businesses began to cut back. The stockyards where my dad worked managed to keep most of its workers until 1931. Banks had begun to close in different parts of the country, causing panic, as bank deposits were not then secured by the Federal government. Individuals simply lost their money if their bank was one that closed. The Great Depression covered quite a span of years, lasting up to 1940 and the beginning of World War II. I grew from childhood to adolescence in this era of economic deprivation. But having never experienced anything different, I was perfectly content with the little material means we had. It was merely the way of life we lived.

Leaving California for Colorado

When the Great Depression hit southern California, the stockyards were also negatively affected. All the businesses around them also began to lose customers or to collapse. It was like a chain reaction with one business collapse affecting another, and one bank panic spreading to the next. My parents felt desperate. The only solution Mother could figure was to leave as quickly as possible and go back to the Colorado homestead. She went to the bank to take out their money, and the banker desperately pleaded with her to not withdraw it. She said she had to because they were leaving the state. People were panicking and there was a run on the bank. The very next day that bank closed. What would have been our fate if Mother had gone even one day later? It was March, 1931.

Returning to Mother's Homestead, 1931

My parents—William Marion Black, then sixty-six, and Letitia Brunia Black, forty-three—had an old blue Chevy flat-bed truck. Being almost two and a half, I have a few memories of my parents packing their food and belongings as they organized things for camping in the back during the long ten-day trip to Mother's ranch. Sturdy side-boards were constructed to hold their belongings, over which they stretched a large canvas, elevated in the center for standing room. I remember the chairs (one still survives) and other furniture stacked tightly high against the sides. To leave nothing behind, Mother had butchered and canned all their chickens. (Dad got so tired of chicken meat every day that afterward he would eat only the gizzards.) After one final check around and making sure everything on the truck was securely tied down, we were Colorado bound.

All the preparations gave me a glow of excitement. We three rode in the cab, me in the middle. Dad did all the driving. Mother was the navigator, and to keep Dad alert elaborated on the scenery of mountains and deserts along the route. We traveled and camped in the truck, all sleeping together in the back under a layer of quilts, which made me feel snug and safe. Roads being as they were back then, travel routes didn't have many short-cuts, so the trip may have covered more than 1200 miles. The old 1920 model Chevy flat-bed truck averaged 25 miles per hour or less. The folks pulled off the road before dark to camp and prepare our meal over a little camp stove. If nights were likely to freeze, Dad drained the water from the radiator. Extra water was always carried in 2-gallon canvas bags hung on the side mirror.

The last, but terrible, leg of the trip past Walsenburg is engraved in my memory by my parent's struggles over a nearly impassible road—a mere wagon trail that followed the fenced section lines of homesteads (the owners having failed and moved on), snaked around, heading in directions where arroyos were able to best be crossed. I remember deep-sinking mud from the melted snow that had saturated the prairie soil, and the road, so seldom traveled, was only slightly packed with deep ruts. The truck would lurch, skid, and then become mired down. Mother and Dad shoveled it out time and again, then searched for weeds, cactus wood, top sod, anything to put in front of the tires for traction. It took three days to get the thirty-five miles from Walsenburg to the ranch, at which we arrived late in the afternoon of the third day. They were so near home, it was frustrating. The last night before the ranch they were within five miles, near a little butte we called the Hogback, again bogged down. The next morning after a hasty breakfast, I watched Mother and Dad from the cab bringing brush and rocks from the side of the butte to put under the wheels until they got dug out. It took seven more hours to make the last five miles.

Arriving at the Ranch

It was late afternoon on a cold and blustery March day when we arrived. The first thing after getting out of the cab was to hurry inside the house and see how the place had fared in the folks' absence. I can't recall if the windows and doors were intact, but I sure remember the floor. The pack rats had moved in, such a perfect residence, safe and dry from the elements. They had heaped up their nests much higher than I was tall with dry cow pies, pieces of dry thorny cactus joints and old gray cactus wood, bits of lumber, old tin cans, and small bones. It was an amazing sight, this my introduction to home. It was the typical fate of a homestead cabin when left unused.

Just as we stood eyeing the pack rats' work the wind picked up and a tremendous gust ripped the canvas off the truck. This was the last straw, yet my parents counted their blessings that the canvas had kept intact for the entire trip. Mother had full intentions to sleep in the truck another night, doubly so after seeing the condition inside the house. She wanted to shovel out the debris and sweep up before we brought things in from the truck, but now that wasn't to be. With the wind blowing a gale, black clouds were rolling in fast from the west and threatening to rain; things had to be brought in quickly off the truck. I wanted to run around after being cooped up for days, but Mother put me in the cab out of the way and safe from rattlesnakes. That night we moved in with the pack rats, hoping there were no snakes inside their huge nests.

In the morning, Mother began removing the pack rat rubble as I watched, curious what all might be inside the nests. A pack rat that darted for safety barely missed getting a whack from Mother's shovel. She swept, cleaned, and set up our scant furnishings as Dad moved heavy things. He set up our bedroom: my cot along the south side, the head of their bed to the north, a three drawer dresser on the west side with the bottom drawer for my things, middle for Mother, and top drawer for Dad. Next he arranged some orange crates for shelves, and the "closet" at the end was just a six foot rod to hang clothes on. I heard Mother muttering about neighbors having come and taken the kitchen cabinet, wood cook stove, and other things she planned to use when we got back. She thought the people that took them should return them, but they hadn't so far. A few days later the kitchen cabinet was retrieved and set up in the kitchen. Mother hammered and fixed to put it back together while I stayed underfoot, helping of course, by handing her nails. It would have to do, even if it was no longer in very good shape.

Two days later, the neighbors came and brought food for a picnic as a welcome back. Mother looked out the kitchen window and was surprised to see a procession of old Fords and Packards arriving. She left me in the kitchen to put on a dress as she hurried outside. We had a little white cabinet with two shelves, about 30 inches high, that her brother Will had made for Mother in Iowa. In it she kept towels, household items, and my few clothes. I stood by it sobbing, as I couldn't put on my dress. I was embarrassed to be seen without a dress, plus scared being left alone with strangers coming. I was by the little cabinet when Blanche Bennett came into the house and knelt down and asked me what was the matter. She kindly helped me with the dress. I can't remember who the other people were that came that day, probably the Hall, Nelson, Krieder, and Knowlton families, or the foods they brought. My folks were pleased to have the neighbors welcome them back, with all their joking, laughter, and talking combined with scrumptious food. Ranch women are great cooks.

At the welcome home picnic someone knew who had our wood cook stove, so a trip was made to bring it home, for the borrower hadn't and it was critically needed. At last food could be cooked, water heated, and bread baked. Mother had a yeast starter that she had kept alive from her homestead days, having brought it in her trunk from Iowa. I would stand close by to watch her kneading bread. There was nothing quite like the aroma of fresh baked loaves when they came out of the oven later in the day. Mother performed magic with that old wood cook stove.

We always had food on the table. What was brought from California was supplemented by edible weeds such as lamb's quarter and new tender Russian thistles along with meat from rabbits Dad shot. As soon as the ground dried Mother would begin planting the garden.

Beginnings in Colorado

I don't know how much money the folks had left, likely very little. A ten-dollar bill was tucked in the Bible, giving a feeling of security to never be completely broke. With the house livable, the folks turned to repairs on the outside, which required hard work with what materials were found lying around.

The old gray barn still stood, but it was missing many sheets of its corrugated tin roof. The folks searched the prairie in directions the winds had most likely blown the sheets of roofing. The pieces they found were drug home, then they hammered the bent tin straight enough to nail back onto the roof. Heavy rocks were lugged up the ladder to weigh down the edges to prevent the wind from blowing it off again. Broken boards and timbers had to be nailed back together. No nails could be wasted; those bent were hammered straight enough on a rock to use again.

The cellar

The underground cellar wasn't in too bad a shape. Water had flooded down, and I remember my folks carried out many bucketfuls of silt deposit. The heavy cedar roof timbers, however, remained strong. The cellar had a domed sod roof like the cabin; more soil was piled on top for both insulation and to prevent leaking. Luckily, the sandstone slab walls were still intact, but the steps down needed fixing.

The windmill was in trouble. During a flood the huge rock that held down the platform on top of the well had fallen into the well. I watched as my parents struggled mightily with only a block and tackle and ropes to get this slippery, likely 300 pound, boulder up twenty five feet from the bottom platform, just above the water level, where it had landed. The gearbox at the top of the tower needed grease. New packing (from a thin cotton rope) also needed to be greased and tamped with a screwdriver around the ramrod, and the water pipes put back in place. With the ever present wind turning the huge wheel, a steady stream of water soon began filling the tank.

In places some of the picket corrals had fallen down and needed wiring back up and broken pickets replaced. In those years ranchers made their corrals out of cedar pickets (the straighter limbs cut off when fence posts were made from cedar logs). They were wired very tightly together top, middle, and bottom so no animal could squeeze through. The pickets were generally six feet high. The Sanford ranch corrals for wild horses had pickets some twelve feet high and were circular to deter jumping out.

The outhouse had toppled over. Mother and Dad dug a new seven-foot deep pit to set it over. They used the dirt from the new hole to fill in the old one. Our outhouse bench had one hole, and on the end was placed an old Montgomery Ward catalog to use for paper. A pail of lime was kept on the floor; a little scattered down the hole eliminated the smell and deterred flies. Later, Mother planted a thicket of tamarack, a tall, graceful, pink-flowering bush, on the house side to screen the outhouse from view.

Two trees near the house had survived. With hot summer sun a shade tree is very welcome. One was an old cottonwood tree. It wasn't a big, magnificent tree, and it had some dead branches, yet it had the honor of being the tallest and only shade tree in our yard for many years. I climbed the cottonwood tree when I got older. Its leaves could be folded in half, drawn taut, and blown through to make a whistle. Wiggling the bottom with a finger created a tremolo. I blew many cottonwood leaf whistles over the years. In the summer, I often lay down on the ground in the shade of the cottonwood and looked up through its leaves and branches at the blue sky overhead. This tree was definitely a part of my early life.

The other tree was a small gray-leaved Russian Olive that had withstood the drought. It had little yellow flowers once in a while and hard green olives that were far from edible. I liked it because it was always there, but it was thorny and not good to play with.

By the corner of the cabin a thorny, small rose bush had survived. When summer came it was covered with single, nicely perfumed yellow roses. Near the yard gate the woodbine had also endured.

Homestead Cabins Didn't Have Running Water or Bathrooms

In this land of scant rainfall, homestead cabins didn't have the luxury of running water. Many homesteaders who didn't have windmills had to haul their water from a distance with team and wagon. Except for the homesteads along Saunders Arroyo, where underground water was usually found, few homesteaders had a windmill. As underground water was rare, and could as likely be under a knoll as in a more auspicious draw, if it was on the homestead at all, a water witch, or dowser, was hired to search for water. The best dowser in our area was Dave Fosdick. The dowser held a small forked branch cut like a "Y" with both hands, often a peach cutting, and when the neck of the rod turned down, despite holding it straight, it was assumed to be a likely spot to dig a well. Even so, many dry wells were dug.

The outhouse was the toilet, with a Montgomery Ward or Sears & Roebuck catalog for paper. People were never excessive with tearing sheets from the catalog, knowing the pages needed to last until next year's new catalog arrived and the old one taken to the toilet. One usually spent as little time as possible out there, especially in winter when Colorado temperatures frequently dropped below zero. Also, the first one out after a snow had to break a trail and shovel snow to get the door open. The worst thing that ever happened at our outhouse occurred one morning when Dad was bitten by a black widow spider while he was sitting on the hole. It affected his nervous system quickly and severely. That was the only time a doctor was ever summoned. Mother went ten miles to the closest neighbor with a telephone. It took Dad several months to fully recuperate. We became more careful to look for spiders after that.

There were a few years when I was quite young that I was terrified of the dark. On the last trip out to the outhouse before bedtime on moonless nights it would be nearly pitch dark. I would run like the wind to the outhouse and back to the house before anything got me. Mother tried her best to relieve this fear, but it took several years before I felt safety from the darkness rather than fear. For nighttime we kept a gallon coffee can as a chamber pot in the bedroom, which Mother emptied each morning. When working out on the prairie, "paper" for toiletry purposes consisted of grass, weed leaves, or pieces of rock. I never saw anyone leave prepared with a catalog page stuffed in their pocket.

For washing our hands, face, and teeth, a wooden orange crate set on end served as the vanity. In those years orange boxes were longer than now and had a partition in the middle that worked fine as a shelf. On top was placed an enamel wash basin (which frequently had a skim of ice on top of the water on winter mornings) and a bar of soap. Above the crate hung the mirror, and nearby stood the water bucket with a dipper. (The water bucket was also used for cooking, dish washing, and holding drinking water. Everyone, even company, drank from the same dipper when they wanted a drink of water outside of meal times. During meals cups were used.) We had a towel rack for the hand towel and washcloth. At the bottom of the "wash stand" was another

bucket to pour in waste water. But it wasn't wasted, because it was used to water plants or for some other beneficial use.

Saturday was usually the weekly bath time. If we were lucky and it had rained, the rainwater caught off the roof was warmed for the bath water. Otherwise water was bailed from the cistern, which stored our drinking, bathing, laundry, and cooking water for the whole year. The wood cook stove was stoked up and bath water put on to warm. The big galvanized tub was placed in front of the stove to bathe in once the water was heated. I was given the first bath, then Mother, and lastly Dad. For Dad's privacy, on cold winter nights, I recall being asked to go out of sight behind the kitchen stove. On honor I never peeked out. In spite of no bathroom, privacy was always kept.

Barbara Ann
By Letitia Black

You are the idol of my heart
From all others set apart.
Child of nature you are blessed
With a fund of happiness.
In the open spaces you belong
With your heart so full of song.
You trip along on merry feet,
Scarcely taking time to eat.
But I'm sure that you would stop
Right in the middle of a hop,
To partake, to partake,
Of a piece of chocolate cake.

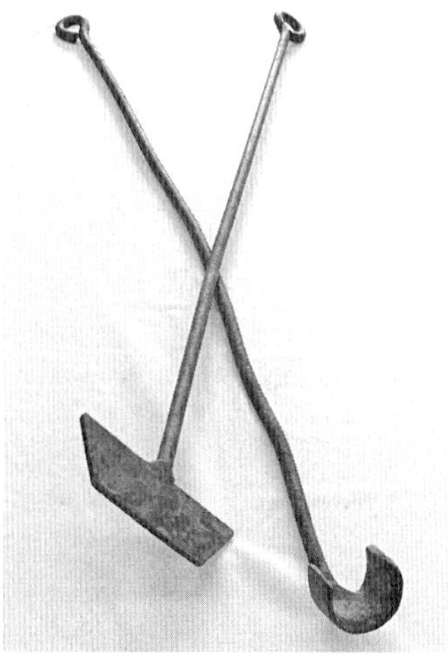

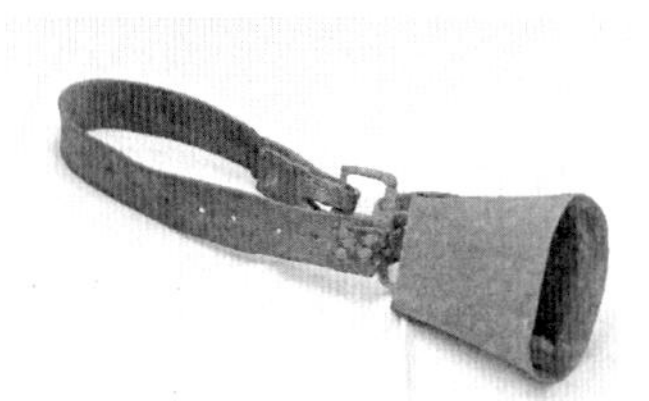

The 3N brand and branding irons. We had two of each, a half circle to make the reverse three and a bar for the N. Also shown is a cow bell. The bell cow appeared as if she enjoyed her bell. Mother dearly loved the new born spring calves.

Part II

Beginning Again

The first start of livestock after my parents started over was a small bunch of sheep.

Pictured is the “T” sheep branding iron for the paint and some sheep bells.

Starting Over from Scratch and Flat Broke

Having gotten the house livable and the outside structures somewhat usable as fast as possible, Mother and Dad began contemplating a way to start making money without any capital. Having the land, they considered pasturing either sheep or cattle for someone else to be the most feasible and fastest means. Their goal was to get a start of sheep and cattle of their own.

No act of nature or the economy of that era could daunt my mother. The land was hers, and in spite of every adversity she was determined to make a living and keep healthful meals on our table. It would take a variety of livestock, garden, and loads of ingenuity, with the accompanying hard work needed, to survive the combination of Depression, Drought, and Dust Bowl years. She wrote to an Iowa friend at the time: "It's quite a struggle to keep the wolf from the door."

Our Garden

As soon as spring weather began to warm the soil, the first immediate step was to get a garden started and producing as quickly as possible. Mother had brought what few seeds she had from California, a shovel, a rake, and a hoe. Great care had to be taken that all the seeds, if possible, would sprout and live. The garden planting was a source of heartbreak and tears for me. I was crushed, because I didn't see myself as a child, but a peer member of our family of three. I told Mother, "I want to hepps you, at least hand you the seeds," but she said I had to stay out of the garden. She explained that my small footsteps could keep seeds from coming up. I was to stay out by the edge. No crying or pleading changed her mind. Mother was pure Frisian with high blushed cheek bones, tall and lean. She was too proud to ever accept unearned charity. This garden plot had to be successful, and it must sustain us for months to come.

The garden area was fenced against livestock intrusion, except where it bordered the corral, which had chicken wire attached lower down. As evidenced by the arroyo banks, the soil was a few feet deep, non-rocky, fine, gray, and fertile, especially with the aid of manure and water pumped by the windmill. The water storage tank was placed on a higher level than the garden, which sloped gently downhill toward the arroyo. Spading and preparing the soil was done one strip at a time in order to plant while the soil was still moist. Mother made a string line with two pointed sticks to hoe straight rows, then crouched down to position the seeds, carefully not sowing too thickly to conserve the few she had, and to eliminate the need for thinning later on. Once the seeds were placed, she lightly tamped the soil taken from the row back over them to the depth each kind required. For watering, a shallow trench was dug beside the seed row and a hose placed at the top of the furrow to allow gravity to take water slowly to the bottom. Dry water rows were hoed shut to prevent evaporation and stop weeds. In the following years, after we had sheep and cattle, manure was spread over the garden before being spaded. With the hot Colorado summer sun, Mother had a bountiful garden.

In 1905, Luther Burbank developed the sunberry, a relative of the tomato, which grew up to 18 inches tall. Not only was it prolific, producing clusters of pea-sized, luscious tasting, dark purple berries, but it was tolerant of harsh winters, arid summers, and poor soil—a fruit homesteaders could grow in place of fruit trees and berry bushes. Mother had planted sunberries when they first homesteaded, brought her seeds to California, and now back to the ranch. A row was planted, saving several seeds in case of a mishap. Sunberries were delicious raw with a little cream, but Mother's sunberry pies were even more famous than her apple pie.

Every day the garden was carefully checked for dryness, weeds, insect work, or animal damage. One morning when some plants were an inch high, she noticed that several were nipped off. Mother assumed it was the work of rabbits. She found old boards, and with a saw and hammer made three box traps for the garden. She lightly propped the boxes up with a stick and placed something she hoped rabbits would like around the base of each prop. If a rabbit bumped the prop, the box would fall over it. To save the rest of the plants she lightly covered them for the night with cloth. The next morning one of the boxes had fallen with a rabbit still underneath it. The twenty-two was brought out, and we had a treat of fried rabbit that day. Mother made sure each trap was carefully set each evening, and there was no further rabbit damage, while we frequently enjoyed their delicious meat. She knew of tularemia in rabbits, so she carefully examined each liver before they were cooked.

The next garden threat was some tiny black insects that jumped like fleas. Mother had a cure for them too. She put a small layer of kerosene in a pan of water, went out early while it was still cold before the fleas were active, and gave the plants a light tap while holding the pan underneath for them to fall in. This worked well for other insects that arrived later in the year, such as striped cucumber beetles. Some insects are deterred by a light sprinkle of wood ashes by the plants. For those that like to hide under something during the night, dry cow pies were placed in the garden, especially under squash plants. Early in the morning, before the beetles came out from their refuge, the cow pies were lifted up and any insects present were stepped on. Striped dill worms (that turned into tiger swallow-tailed butterflies), fat tomato worms (that became lovely Sphinx moths), and any other bigger worms or beetles were searched out and picked off by hand. Aphids would get on the cabbage, and if there were too many for the lady bugs to handle, they were carefully mashed. No poisons were ever used on our garden, as Mother figured out alternative solutions. The greatest threat of all to the garden was the summer hail storms. For those there was no cure, no way to prepare.

During our first year back at the ranch, when a neighbor divided her rhubarb patch, she gave us a bag of crowns, which Mother planted in a long row on the corral side of the garden, where the westerly winds blew fine manure dust into the garden. Dad loved rhubarb. He called it the pie plant. Asparagus (which Dad called spare grass) grew wild along the irrigation ditch banks in the Arkansas Valley. Jeffery's family brought us a box of roots, which Mother planted deeply several feet behind the rhubarb row. Another year we received some Jerusalem artichoke tubers (a sunflower relative), which were planted behind the asparagus. Mother used these multiplying tubers raw for salad or cooked in place of potatoes. Once established, these three perennial crops produced in abundance year after year in this very fertile location. Rhubarb and asparagus were our first garden delicacies of the spring, and the artichoke tubers were used later in winter salads.

Mother saved seeds from the garden plants. Near the end of the season seeds were allowed to mature in such plants as beans, peas, corn, spinach, and lettuce. Some plants had to be protected over winter in a deep layer of straw for seeds, like beets, carrots, parsnips, and much of the cabbage family, which produce their seeds the second year. For melons, cucumbers, squash, tomatoes, and pepper type plants, she saved seeds from the nicest ones we ate and spread them out to dry. Few seeds were purchased, and certainly never hybrid seeds that could not reproduce their own kind, as are sold nowadays. Seeds were often mixed with other small particles and needed to be winnowed. This process involves two pans and a very slight breeze. The pan with the seeds is held high and poured slowly into the lower pan so that the breeze carries away the unwanted chaff. This process is repeated a couple times until the seeds are cleaned. Mother then put her new seeds safely in small bottles and jars, and many in Prince Albert tobacco cans given to her, to protect them from mice and moths, and labeled them. Seeds were very precious to Mother.

A Start of Sheep

When the folks came back to Mother's homestead at the beginning of the Great Depression, a few head of horses were the only livestock that remained. For our livelihood, a new start had to be made with sheep and cattle. Mr. Stauder, whose sheep ranch Dad had managed before he and Mother went to California, offered to have us pasture some old cull ewes. In return, he was to have the wether lambs to sell, and some ewe lambs. We would keep the old ewes and a few of the ewe lambs to start our herd. It was a barter without cash. At least we had land to graze the sheep, such as it was in those dry years. Sheep could not be left alone to run off, get lost, or killed by coyotes. One parent always tended the sheep, but could not leave a child alone at home, so at three and four I tagged along. Mother had many concerns. There were times she'd have to walk fast with her long, quick, agile stride, or run, to turn the sheep. Sometimes she yelled "woof woof" hoping the sheep would turn back. She'd try her best to make me stay, saying "I'll be right back," but I would never be left. With an inner need to be capable, I insisted on running with her. She said I was determined. The presence of rattlesnakes was a further concern.

I have many cherished memories of herding sheep with Mother. She knew to make valuable use of time, especially when the sheep were calmly grazing. She sang me old-time songs like Nellie Gray

Herding sheep with Mother.

or Old Folks at Home. There were favorite stories, such as Little Black Sambo and Epaminondas, I'd ask to hear many times over. Her dramatic acting of the characters and voice imitations made me laugh. Walking at her side was my first schooling in spelling and arithmetic. Mother used this time out on the expanse of prairie to instill within her child an affinity for nature and its myriad forms of beauty. She brought to my attention lovely sky colors, a gold rim around a cloud, sometimes a silver lining. Cloud formations that first resembled a wolf, then with atmospheric winds plus imagination soon changed to flying geese. She pointed out delicate prairie flowers by name: dainty pink phlox, early white lilies, tiny yellow violets. Attempting to describe the Creator to me, she said that the human soul was even more wonderful than these wonders of nature.

Lunches were brought of homemade bread and peanut butter, hard boiled eggs, and a rare treat of an orange if we'd been to town. I especially loved to eat at an exciting area we called the City Rocks whenever Mother grazed the sheep there. She spurred my fantasy in this vicinity, where there were scattered dozens of huge sculptured sandstone boulders and slabs of rock, all shaped differently by centuries of the elements. She called them stores and buildings, while I filled in with make-believe people living and working in them. I also had splendid times climbing and playing among these gigantic outcrops. One monster boulder had a three foot wide crack sixty feet long. I loved to go through this tunnel time and again, even though it was windy, dark and scary. Mother always checked first to be sure there were no snakes.

When we grazed sheep in the Barry (as we termed Will Barry's section), we passed the remains of deep test holes drilled to check for oil. Two huge spools that had once held cable had been left behind. I never tired of climbing and pretending on the wooden spools, which were about three feet high and four or more feet wide.

On zero temperature days, sometimes we'd be at a location where Mother could build a comforting small fire below the bank of a little gully and out of the chilling wind. I would hold my hands, palms out, toward its warmth, feel its heat on my back, gaze into the flicker of flames, and all the while Mother was searching to find a few more pieces of cactus wood to keep it going. Sometimes we ate our lunch near the tiny fire. Often the snow lay on the ground for several weeks with daytime temperatures ranging around zero. This never meant that we could stay indoors. The sheep needed to be taken out to find whatever grass showed above the snow. Mother was concerned that my small body would get frost bitten or even hypothermia, so she always kept me walking, in fact, mostly trotting to match her long stride and brisk pace. She had other means of keeping me warm. One special way was a game she called "Fox and Goose". She would tramp down a huge circle in the crunchy snow. Inside the circle she stomped the snow down to make spokes or inner trails. Then Mother would pretend she was the big bad fox and try to catch me, who was to be the goose. She kept me running and running and laughing as she faked attempting to capture me. Never once did I suspect we played for anything but the fun of the game. It was only years later that I realized it was a way of protecting me in freezing weather. My wise mother solved each adversity she faced one way or another.

Our little herd of sheep began to increase by raising bum lambs. It was not long after we began pasturing the sheep that my parents acquired twenty starving dairy cows. (More on that later.) Dad would drive out to big sheep ranches and ask for their unwanted lambs: a weak twin, orphans, ones left to die, which the ranchers were happy to give away. That spring Dad found over twenty bum lambs for us to bottle feed on the dairy cows' milk. Even I could hold a bottle with both hands, one on the top of the nipple so the lamb didn't jerk it off. It was amusing to see their tails wiggle in delight as their tummies began to fill from just a hollow spot behind the rib cage to being pushed out. These lambs were kept in the barn or corral while we grazed the sheep herd, until they were strong enough to come along, following us as their mother. Three were my special pets, whom I named Blackie, Brownie, and Whitie. They became my playmates, and they stayed close like three shadows at my heels. A year or more later when I herded sheep on my own, two bum lambs, my companions Rosemary and Lilly May, had grown to maturity. I loved to ride on their soft, woolly backs, until the folks happen to notice they were getting sway-backed like an old horse!

One evening Mother counted a dozen sheep short. Evidently a little bunch had wandered off when the sheep had been left alone, no one being able to watch them at the time. Having no idea where else to look after all possibilities on the ranch had been exhausted, Mother said we needed to track them by following the fence line to find out where the tracks went. Mother searched part of the fence line and Dad the other. Neither of them found tracks, necessitating a second round of careful searching. It turned out that Dad had not been thorough, and the sheep had gotten out to the northeast into the Sanford range. By that time it was getting too dark to search, but I remember Mother being pretty disgusted with Dad for not checking every foot of ground.

The next morning at dawn Dad and I started the search. We had gone several miles before finding the first sheep. Two were standing on the top ledge of a butte of the Sanford hills, evidence that the sheep had been chased by coyotes during the night, and these had escaped, running to the butte top for safety. We did not disturb or scare them, planning to get them when we returned, hopefully with other sheep. Dad and I searched another three miles to the other side of another range of the Sanford hills, where a hogback of jagged basaltic peaks protrudes from the prairie. There we found three slaughtered sheep and two alive. There was no telling where the others would be, if even alive. We herded the two jittery sheep toward home, picking up those on the butte top.

Spring is lambing time. Of our seventy head of old ewes, some did not have a lamb. Personal care was given each sheep that "dropped" a lamb, putting ewe and baby in a small panel pen in the barn for a couple days to make sure they bonded and the lamb was healthy and sucking. If a lamb was too weak to suck, Mother crawled in the pen and pinned the struggling ewe against a side with her shoulder and knee while holding the lamb up to nurse, often putting the teat in its mouth to get it started. If very weak, she milked the sheep and gave her milk by bottle to the lamb, always making sure it was only her milk fed to her lamb. Sheep accept their baby by smell, not only of their body, but its droppings as well. Adoption was tried on ewes whose lamb had died. It takes patience and time. The lamb to be adopted would be rubbed over the ewe, especially her hind end, the sooner after she had given birth the better, to give her scent to the adoptive lamb.

One ewe we called Big Teat Mama had a very large lamb she loved dearly that died. She put her head just above the baby giving loving bleats, and she pawed to get her baby up, holding her leg back and nudging it to nurse, but it was of no avail. That night, the next day, and the following morning, she continued to snuggle against her baby or stand over it trying to get it up. The next day a twin was born needing adoption, and we put the dead baby's pelt on her. Big Teat Mama was skeptical if it was really her baby, but in time she claimed it.

Before letting a pair out of their pen, Mother "branded" them. She made numbers zero through nine out of heavy wire, leaving a stem for the handle. With linseed oil and black stove soot (also called lampblack), she made a paint that didn't wash off with rain. Each pair was numbered alike before being put into small groups.

Lambs have an ordeal to go through. They are caught, usually by grabbing a hind leg, and drug over to where they are to be docked. We gathered a hind and front leg in one hand, and the other back and front leg in the other hand, and set the lamb on its rear end on a ledge so its tail hung over for Dad to grab and with his razor sharp pocket knife quickly cut it off. Sheep are given an ear mark; ours was the tip cut off the right ear and then a half inch slit in the middle of that tip. Male lambs had more to suffer, because they were "cut", as we called it, or neutered. Then it was set down and the next victim was grabbed. There was considerable "baaing" on docking day from concerned mothers hearing cries from their lambs.

A couple years after getting sheep, the government paid the ranchers to kill older livestock. This was supposedly to keep the prices up, and also because there was not much feed in drought years. For this project, Dad did the killing work, using a sledge hammer to give an old ewe a quick, hard blow on top of the head, which dropped her, stunned. He then quickly cut the jugular vein. When kicking ceased, and her body was still warm, he skinned off the pelt with his one and only pocket knife, which he kept sharpened with a little whetstone. I came in handy to help Dad, by holding the leg he was skinning out taunt, and then the others, making skinning easier. Once all the legs were skinned out and the ewe rolled on her back, he made a straight cut down the throat and belly to skin the hide from underneath up the sides. Propped up between his knees, Dad cut the hide loose from the base of the head, gave the pelt downward jerks, and some quick cutting back and forth to loosen a foamy substance between the fleece and body. It was finished, and now for the next one.

The pelts were laid out to dry and cure to be sold. Dad took care to not make any cuts in the pelts. Some pelts were kept for rugs, which felt so good stepping out of bed on icy mornings. Not to waste anything, Mother canned the best of the leg meat, and the chickens pecked meat off the bones. The carcasses were hauled off a mile to, as we termed it, the "bone pile". Steel traps were set there to catch coyotes for their pelts. At that time the state of Colorado was paying a 50 cent bounty on coyotes in an effort to reduce their predation on livestock. Unfortunately, the traps also caught skunks, and the trappers resorted to burying their skunked clothes to rid them of the terrible smell before they could wear them again. The bones eventually bleached in the sun and could be seen from the house as a white spot.

Cousin Jeffery Rescues Me

I loved tadpoles. Every place that had tadpoles was a temptation. With shoes off, and pants rolled up, I'd wade in after tadpoles. One particular time, my cousin Jeffery was visiting, and we were herding sheep by the lower canyon, where there were potholes in the rocks recently filled with water when the arroyo ran. There were many little pools; a couple were shaped like bathtubs. This was where I frequently brought the sheep in the heat of the day to water, then I would lie in the shade of the rock ledges and overhangs on whose ceilings the swallows had attached hundreds of their mud-daubed nests. On this occasion, I waded into a pool whose rock sides were slick and slimy. My feet flew out from under me so quickly, I didn't know what had hit me. I had no idea how to swim, and the sides were so slick that I'd slip right back down, taking in gulps of water and splashing wildly as I went under. Jeffery thought there was a whale in the pool and ran off to hide. But he quickly figured it was me, or at least he saw me still trying to get out and then going under. He pulled me out of the slippery pool. We were each six years old, but I knew he had saved me from drowning. I was so thankful, as it had been a terrifying experience. To this day, I get frightened when my head gets under water. After the drenching, I walked home in the arroyo to keep out of the cool summer wind, clothes soaked and teeth chattering, while Jeffery herded the sheep home.

Sheep Raising

Two hazards for our sheep were poisonous weeds and cockleburs. Larkspur and death camas grew on the prairie, which we pulled up every time a plant was located during sheep grazing. We also pulled up whorled milkweed and cockleburs, which grew in the arroyo and damp draws, but more would always be found the next year. When young, cockleburs resemble a sunflower. The sheep apparently didn't have a taste for them, but in the fall, as the burs matured and turned brown, they stuck like leaches to their wool. These inch long, brown burs, entirely covered by little hooked barbs, were extremely difficult to remove. Wool did not sell with burs in it, so they had to be eradicated as best possible. In the fall the sheep were not allowed to graze near where cockleburs grew, and Dad, Mother, and I took sticks and beat the burs off their stalks, hopefully not missing any for the sheep to get caught in their wool.

In the late spring or early summer, after the lambs were well started, it was shearing time for the sheep. When we had only a small flock, Mother sheared them by hand with a pair of shears that looked like grass clippers. There is a certain way to shear, so the fleece stays in one piece. One starts shearing their legs and neck out first, then under the stomach area and a bit up the sides, next from the shoulders down the back and sides, so that the fleece comes off in one piece. It is then tied with "wool string", a kind that dissolves at wool treatment places. No sheep is going to sit docilely on its hind end between your knees; its struggles and kicking are apt to scatter the fleece that you are trying to keep intact while shearing. After the sheep were sheared, each was given a painted brand, again using Mother's black linseed oil paint. Our brand was a big "T" on the right side, which showed even after the wool began to grow back. Wool was sold in huge gunny sacks, tamped or packed in firmly. Our sheep gave us two cash crops: summer wool and fall lambs.

By the late 1930s our sheep raising had changed. Jeffery's dad acquired land and sheep he ran south of us. Our sheep were merged into his sheep herd; his sheep had a "7UP" brand. Summer and fall sheep grazed on his land, and he took care of having a sheepherder, keeping fifty pound blocks of salt out, stocking his camp wagon with supplies, and keeping it moved frequently so the sheep didn't graze in one area too long. Our land was used prior to lambing time until after the herd was sheared. With many more sheep, lambing time required much work. Throughout high school Jeffery and I left six weeks before it was over to help at home, mostly to take care of several little flocks of ewes and lambs by herding them on horseback. One foggy morning, I was watching three little bunches, each some distance apart, on Ginger. Soon the fog became so dense I could scarcely see any distance. Ginger refused to go in the direction I thought was toward home. When the fog lessened, I was most surprised to find out she was right, and that I had been lost.

Mother spent long hours to keep the lamb mortality rate low. Periods of cold drizzly rain were especially bad for weak newborns. To keep them from becoming chilled, I would bring them, wrapped in a gunny sack, to Mother, who kept the kitchen stove hot with the oven door down. If a lamb was quite "far gone", she'd pry open its cold, clamped jaws and insert a nipple for warm milk.

Shearing time was now far different. A Spanish-speaking shearing outfit, often from Texas, would come to all the sheep ranchers to shear their herds. They were a spectacular sight to cousin Jeffery and me. Some twenty men would come in with their trucks and equipment. They first set up a framework with an awning to provide cover for eight to ten men working at shearing stations on either side, depending on the size of the outfit. The ground area was covered with plywood to keep

Men shearing the sheep.

the wool clean and hold down the dust. Their shearing apparatus was powered by a huge gasoline motor with a square metal pot attached on top to hold water for coolant. Down the center of the framework near the top ran a long shaft which powered all the shearing stations. Each shearer had his bucket of spare sharp shearing blades, his oil can, rags, tools, and wool string. A bunch of sheep without lambs would be corralled around this shearing station. When a man finished shearing a sheep, he tossed the tied fleece outside the shearing area to be bagged before catching the next sheep. When the last sheep was sheared, they were let out and a new bunch was brought in.

Jeffery and I were assigned the chore of bagging the wool. A tower had been built to hold the burlap wool bags in place while they were being filled. The bags, which were eight feet tall by three feet wide, were way over our heads, and when empty, one couldn't get out from inside alone. One of us tossed up the fleeces, while the other got down

Filling the wool sack.

into the bag after a number of fleeces were inside, and began jumping up and down to get the wool tamped in as tight as possible until the bag bulged. The sack was let down and sewed shut with a long sack needle, leaving a six inch "ear" on both ends. Then on to the next bag, and the next. It got very hot in the bag (a good thing we took turns), and we became well oiled with lanolin and smelled just like sheep. A bit of fun afterward was to jump around on the stack of these big bouncy wool bags.

Branding the sheep.

After the shearing was over, the next task was to get the sheep branded. Our sheep and Jeff's had been separated for shearing and to keep the wool separate. After a year's worth of wool growth, a brand can become quite obscure, but at least the earmark is a sure sign. A long, narrow chute was built along the back side of the barn for the sheep to pass through single file. With the sheep crowded in, a "T" when ours, or for Jeff's a "7UP", was branded with linseed oil paint on the side of each sheep. Then the next group was put into the chute. Now the herd was ready to go out to range for the rest of the summer and early fall.

You've heard the expression "follow like sheep". No sheep wants to be first, definitely in a strange situation, like going down a chute. One can barely push or drag one out of the herd to go first, but once one goes, the rest all follow eagerly. Sometimes a goat is kept with the sheep for the purpose of being a leader. When the sheep band was large we had four black sheep as markers. If one was missing, likely a bunch of sheep would be gone. One time a black sheep turned up missing, so we assumed some of the rest of the sheep had strayed off with her. We hunted everywhere. Finally, Mother found just the black sheep stuck in some rocks. None of the other sheep had been lost after all.

As our sheep herd increased and then combined with Jeffery's parents' sheep, so did the number of rams. At one time we had six huge Rambouillet (a breed well adapted to Western range land, excellent for both wool and meat) with gigantic curving horns. They would gang together and had horrible fights, usually in pairs. The two contestants would back far off from each other; if they didn't think it far enough, they backed still more. Then with a burst of speed they ran at each other, heads down, and slammed their skulls together. The term "thick skulled" well-applied to these rams, for they never stunned or killed each other, although they did shake their heads afterward. One round was never enough; more like four were needed

Rams butting heads

before both gave up. I couldn't help but feel frightened when herding the sheep alone for fear they might attack me. I made plans how to save myself, but nothing ever happened; the rams ignored me.

Drunk Sheepherders

By combining the two herds of sheep into one band, it was large enough for our parents to afford having a sheepherder. I don't recall how we divided the cost of the sheepherder, but one of Jeff's responsibilities was finding the sheepherder and taking him to or from town as needed. When a sheepherder got the "urge" to go into town, nothing could deter him; he'd usually have a severe toothache, some "valid" reason to go in for a couple weeks. This particular time Jeff had brought Vicente back to the ranch while still under the influence of his town binge. That afternoon I was out jumping and playing on the tumbled-down brick chimney that had been the fireplace in Aunt Bertha's homestead cabin. I happened to see Vicente lying on the ground nearby, so I went over to him and discovered that he was dead. His mouth was open and big blow flies were buzzing around inside. Frightened to see a dead man, and sad, too, as I liked Vicente (he had tried to teach Little Jeff and me how to speak Spanish), I ran as fast as I could to the house to tell Mother. She was startled and hurried out to where I saw him. She laughed and told me he wasn't dead, just knocked out from drinking.

Les (Telesforo Valdez), the large, gangly Aztec-Mexican with two front fangs for teeth, acted differently when he was hauled back to the ranch from town, after he "had to go see the doctor." He never slept his drinking binges off, or at least I didn't see him sleeping one off. He would be very obstreperous, and more than once, if Little Jeff or I were around, thinking of visiting him, he picked up the ax from the woodpile, held it overhead, and came running at us, yelling, "I kill you!" "I kill you!" We had to run like the wind to keep out of the way of his flailing ax, because Les could move his normally shuffling feet really fast when he was drunk. In a day or so, he would be back to normal. We didn't bring up the subject, nor did he, so I don't know if he remembered what he had done.

One of our sheepherders believed in brujas (witches). He had seen a bruja at night lately carrying a lantern back and forth, back and forth. He was scared and ready to quit. An airplane beacon had been installed on the north side of Rattlesnake Butte. It was a new thing out in this range country, and it was hard to explain it to this frightened, superstitious man.

Black Grama Grass
By Letitia Black

Black grama with your gallant flag,
I doff my hat, and thank my God
That he made you, that can withstand
The fierce onslaught of wind and hail
The searing blast, the icy gale,
Then, with the spring's returning rain
Grows green and soft and lush again

Dairy Cows and Dairying

In the Depression years many farmers could no longer afford to feed their cattle, which happened to one farmer with twenty milk cows in our area. The banker informed my Dad of the situation and told Dad he would lend him the money to buy the herd for seventeen dollars a head. Dad went and looked at the nearly starved cows in the farmer's corral, which were weak and wobbling and looked like hides thrown over skeletons. Not sure how much longer they would live, Dad said he'd take all that could stand up when they were hauled to our ranch. As luck would have it, all the cows were able to stay on their feet except for one blue roan we later named Bluebell. The trucker pulled so hard on her tail trying to lift up her rear that her tail broke off about a foot from her body. Once pulled up, even she managed to stand. So we acquired a Short Horn dairy herd, no two cows looking alike, to go along with our small flock of sheep. Grass was scant on our land, but at least they didn't starve, and a few continued giving a bit of milk even in their weak condition. Milk was a delicious addition to our meager diet—milk to drink. Later, when their milk flow picked up, Mother made cottage cheese, butter, and even cheese. During drought years on the prairie, it was estimated that it took thirty-two acres of grassland to graze one cow. We had enough acres to feed the cattle, if only the grass would start to grow again.

Cattle on neighboring ranches also showed every rib, and their hip bones protruded. It was partly because of the land being over-grazed, as the ranchers ran many more head of cattle than their land could accommodate during those dry years. Due to starvation, these cattle started to eat cactus, braving the thorns. Ranchers attempted to burn off the thorns with blow torches, but with so many sections of land, they could never get them all, and their cattle became addicted to eating cactus. Many cows got big cancer sores in their mouths and died because they could no longer eat. Others developed bulging, festering, cankerous eyes from cactus thorns blinding them. Big, bald, red-headed turkey vultures were one prairie denizen that never faced starvation. Flocks of up to a dozen could often be seen circling in the sky before gathering to devour a carcass.

Mother's letter to a friend describes ranch life at the time better than my memory.

Fowler. Colo.
Dec. 22, 1935.
My dear Barbara LueLile
Padgett.
Are you still living? You've most certainly been stingy with letters the past two years, like myself.

I guess you're one of my specials and I want to write lots to you and I put it off till last and before I get thru writing my letters, usually will be too late for Christmas anyway so then I think I'll write a little later, and all thru summer I keep thinking of you. We've been pretty busy keeping the wolf from the door. Coming back practically broke with three years of drought and grasshoppers on top were pretty stiff but a few local rains favored us and we could pasture some stock, and last year sold our old ewes to government for what we paid for them, and this summer had bridge worker boarders for months. Supposed to be 20 but from 12 - 33 mostly Spanish. I got terribly tired of it, but we were very grateful for it too. Really did pretty good with it. Fed them quite a number of wether lambs and had good garden, and canned anything left over: meats, soups, spaghetti, vegetables or fruits, so when they left had over 500 quarts in cellar and buy very little now.

One year marketed 258 turkeys. When the men left about first of Nov. started hauling cactus, 20 big truck loads. A load is a big day's work for the two of us while Barbara watches the sheep. Have our own grinder and engine this year and even ground it ourselves in about the same running time as it took four or five men last year. Put it into trench silo for ensilage, and cattle and sheep are wild about it. Have 18 cows and heifers, and if we don't have boarders in spring will dairy and raise a flock of chickens, and maybe a few turkeys to raise hens for big flock next year. Like especially to raise turkeys for they find the rattlers and circle around them and chatter and we can kill them. That big flock found about 50 for us when we herded them out. I've had a little sticker in my eye for about two weeks, not bad but bothers me at times.

Barbara was seven this month and big and healthy, and very happy. She's reading her second reader at home, we're 10 miles from school. Did you ever make

cheese. I made 3 3# ones yesterday. You buy rennet tablets and directions are with them. We all send best wishes to you & your husband.
Lovingly Letitia

I'm proud of Mother, because she was far more ingenious and protective of our cattle. She must have worn her thinking cap for some while before devising a plan to save them, suitable to what small means we had, likely no more than the farmer who had left these cattle to starve. Our grass was meager, but cactus was profuse. She and Dad dug a huge pit in the ground for making cactus ensilage, then they lined it with straw. They purchased an old, fairly small Letz mill with a conveyor belt for grinding the cactus. The conveyor extended over the edge of the pit to allow the ground cactus to fall inside, necessitating only shoveling to spread it out evenly. When collecting the cactus, they used a machete corn knife for lopping off the leafy branches. With a pitchfork, it was tossed onto our flatbed truck and hauled home to grind and allowed to ferment in the pit. When the pit was finally filled, a layer of straw was put over the ground cactus, and then dirt was piled on top of the straw for insulation to keep in heat to help fermentation. When the ground cactus had fermented, in two or more weeks depending on the internal temperature, we could squeeze the silage in our hands and not feel any stickers. It even had a pleasant smell. Our cows went absolutely crazy over this silage, crowding and shoving each other around the long feed troughs my parents had made in order to get at it.

Our cows began to fatten with this nutritious ensilage and to produce rich, creamy milk. The cows had no idea their feed came from cactus and never attempted to eat it on the range. It was not an easy job for my folks to cut and haul in cactus, make ensilage, then dig it back out of the pit to carry to feed troughs for the cattle. It took weeks and months of work to feed twenty head of cattle until the grass became sufficient again. I recall a time that Mother had cactus chaff in both eyes after a windy day of cutting cactus. They hurt, watered, and tears rolled down her cheeks. She was barely able to hold either eye open. Wouldn't you know, of all times, company showed up. It was fortunate, too, because she asked the lady if she could locate and remove the debris from her eyes. Neither Dad nor I had been able to find the trouble. I recall that Mother, a very hospitable lady, felt bad because she couldn't see to cook a meal for the couple. It took several days for her eyes to recover.

One day I was with Dad when he was hauling cactus. The folks had brought the sheep out to graze where they were cutting the cactus to accomplish two things at once, and Mother had left to take the sheep back to the barn. As Dad moved the truck up from one spot to the next for cactus, I liked to sit on the running board, my legs dangling over the side. Cactus were thickly scattered over the land, so it was no wonder he scrapped the side of the truck into some bushes. The impact drove the thorns deep into both my legs like a pincushion. When we got home Mother was very upset with Dad for ever letting me ride on the outside of the truck. They both worked a long time on my legs with pliers and forceps. The broken-off thorns they couldn't get out remained under the skin on my shins like small round bb's. About sixty years later the last one finally disappeared.

With our new milk cows flourishing, my parents now had the means to make money to pay the banker back. They bought a used cream separator and some five gallon milk cans. The milk separator had a big metal bowl on top to pour in the milk, which first had to be strained through a cloth or filter. Under the bowl was the apparatus for separating milk from cream: a metal cylinder containing forty separating disks with two flat, elongated spouts, one for the cream above the other for the skim milk, and a handle on the side to turn. Dad and I did most of the cranking. At first, it is hard to start turning, until gradually momentum is built up. When sixty revolutions per minute is reached, the spigot at the bottom of the bowl is turned open to allow the milk to run through the whirling disks. The cream is separated and runs into its own can and the milk into another. After every use, the separator was thoroughly washed, the parts scalded, then it was put out in the sun for further sterilization if it was sunny. Mother was the one to wash the innumerable separator parts. The milk bucket and strainer were also thoroughly washed after each use, scalding water poured over them, and put out in the sun for the day. The cream was sold, but we kept the skim milk for our own use. Most of it was given back to the calves and other animals, and some we used for making cottage cheese and cooking.

When a cow first has her calf, the calf is allowed to suck as much milk as it can hold, but as milk cows give far more milk than a calf can use, the rest is milked and given to other animals. The first thick and yellowish milk, which lasts for three to four days, is a very nutritious colostrum. We always left the calf with its mother for over a week to drink all the colostrum before using her milk.

Dad didn't milk, but Mother did, and I attempted to help. As I got a little older and my hands became stronger, I was able to milk ten cows morning and night, and Mother milked the other half. The cows were always milked early in the morning, so they'd have a long day to graze, and they were brought back home for the evening milking.

There was a huge variation in our cows with respect to physical appearance, temperament, and personality. Some had big fat teats, or long, slim ones; others had short, hard-to-grasp teats. Some cows milked easily, giving big streams that filled the bucket, with foam on top, in a hurry. Then there were those that took a hard squeeze to get a tiny stream. Some were calm and stood still, while others were edgy, and you had to watch them every second, as they might sling out a leg and knock over the milk bucket. A few were so undependable about kicking that hobbles needed to be put on their hocks. It's nice when a cow lets down her milk as soon as you start; some will hold it up a long time, especially if a different person milks them, as if you're not their calf and the other person was. Once in a while a cow actually considered me her calf and thought I needed to be scrubbed. That was sweet of her, but their tongue is very rough, like a rasp, and a few licks is plenty. Mother often sang to the cows while she milked them. One of her favorite songs, "Darling Nellie Gray", started like this:

There's a low, green, valley on the old Kentucky shore
Where I've whiled many happy hours away,
Just a sitting and a singing by the little cottage door
Where lived my darling Nellie Gray

The blue roan with the stub tail was the hardest of all to milk. Not only were her teats short, round little nubbins, but it took a strong hand to get her milk. If teats are too short to hold, you need to use your thumb and the next two fingers to strip down and get milk. It was amusing when a visitor would come and saw how easy it looked to milk. They would ask to try, and almost invariably they would get no milk at all, hard as they squeezed. They were perplexed, but there's a knack to the way you squeeze. You have to open your hand on the teat for the milk to come down, then squeeze from the top downward. We usually sat on a wooden box or upside down bucket to milk, holding the milk bucket safely between our legs. We used the word "putty" to ask a cow to put her leg back for milking. The leg had to be moved manually each time until she finally learned to "putty" on her own. A cow's breath often had a pleasant smell, perhaps of sage, and sometimes what was eaten was reflected in the taste of their milk.

There were some annoyances connected to milking, mostly caused by the cows' tails. They wanted to swat at flies stinging their sides, but instead they usually managed to hit me on the cheek. In the spring, when the thistles and the grass begins to green and cattle get the green runs, your face can receive an awful wet, green splat when they give their tail a swish. But fall is worse. That is when the cockleburs get brown and sharp as they ripen in the draws. The brush ends of the cows tails got packed with cockleburs, and a smack with cockleburs on the tail truly hurts. To avert a tail smacking or hoof kicking while milking, I would reach under the cow's belly to shoo off any flies that had gathered there. Also, before one starts to milk, their flanks and the area near where you milk needs to be brushed off so nothing falls into the bucket. Cracked teats were another problem, as naturally it hurts the cow if you squeeze on her sore, cracked skin. A lot of things are causes: a calf has bitten her teats, cuts and scratches from brush or barbed wire, and in winter chaffed skin from the cold weather. A big jar of carbolated Vaseline (similar to bag balm that is sold nowadays) was kept in the barn to massage into teats with sores to keep them soft and to prevent cracking.

The folks ran their dairy and sold cream for about five years before they began to cut back. The cows were eventually bred to Hereford bulls, as my parents end goal was to have a herd of beef cattle. With each generation, the offspring became more heavy set for beef and less suited for producing milk. But those heifers suited to milking continued to be milked. I was probably eleven at this time, and I wanted the exciting job of breaking these strong, young cows to milk. They were wild from living on the range, so they wouldn't allow anyone to get close. I would rope their horns, not their neck, as they'd choke struggling, and then snub them to a corral post. Next, I roped their hind legs and tied them off to keep from being kicked. With that done, I would touch the young, excitable cow, including her teats, until she settled down enough to attempt milking. I took a small bucket to hold with one hand and milked with the other, prepared to jump at any moment. When they gentled down enough that their head did not have to be snubbed close to the post, they were given some slack and a pan of grain to eat while being milked. That made me milk fast to get done before they finished eating. After a while, maybe three weeks to a month, they'd begin to behave like a milk cow, at which time Mother also milked them. Most dairy barns have stanchions to hold milk cows' necks so they can't walk off, but we

Training a cow to milk.

didn't have any. I didn't even know they existed; we just used a rope to tie a cow up as necessary.

At one time the folks had a big, black Jersey milk cow that gave a running-over three gallon bucket of milk twice a day, with an inch of foam on top. One day, she turned up missing when it was time for the evening milking. They looked for her, but it was getting too dark by then to find her. The next morning they went out at daybreak to search for her. They looked everywhere, as it would be very serious to leave her not milked for any length of time. She was finally found way to the west of the ranch, probably lured by some neighboring cattle. Her bag was very tight, and her teats were hard, stiffly sticking out. Her bag and teats were so tight that it hurt to be milked, and she kicked with every squeeze. Mother cut back on feed to lessen her milk and in a few days succeeded in alleviating her swollen bag.

My parents could not afford a bull, and they really didn't have enough cows to warrant one. Fortunately, neighboring bulls from the Sanford and Red Top ranches liked to come court our cows. They were big white-faced Herefords. Some of the Red Top bulls were so huge they looked more like buffalos. I was always afraid of these enormous range bulls, as they would get in terrible fights. They started by bellowing, tossing around their big curved horns in a threatening manner, and furiously pawing the ground, which flung the dirt out behind them or up on their back. They did make a breathtaking spectacle at my age, which I watched in amazement from a safe vantage point. Thus, gradually over the years, our herd of cattle became more Hereford than dairy.

Feeding the Cattle

Just when the weather is the worst, and most people love to be inside by the fire with a good book, is the time that ranchers have to be outside the most taking care of livestock. If the temperature is down around zero, the water freezes up, but the cattle are just as thirsty as ever. The ice has to be chopped in the tanks or thawed out, which takes work and time in the freezing cold.

If there is snow on the ground covering the grass, the cattle need to be fed. During the first few years of cattle raising, the folks couldn't afford to buy hay. But, eventually, they were able to buy a few tons of hay in the fall to store in the barn for when the stock couldn't graze. They also bought several tons of "cake", which is compressed, cubed cottonseed, linseed, and soy. Most everything used to come in hundred pound sacks. A ton of cattle cake would be twenty hundred pound sacks, heavy for either Mother or me to lift alone, so we unloaded and stacked feed together, each taking an end of the sack.

When the cattle were out on the range, cake was hauled out to give each cow a pound or a bit more. But as often as not, when the weather was bad, it was too muddy for a vehicle, so sacks were taken out on horseback in front or in back of the saddle to feed the cows. After getting near, but not too close to be seen or smelled by the cows, and the cake was properly spread out in a long line, I yelled at the top of my lungs, "Sooo Baaa", which is soo boss, for bossie, or cow. The cows had magnificent ears for this call and would come running at top speed for the feed, which is why one had to be ready before calling, or they would gang up around the horse before you had time to spread the cake out. The cows lined up along the row of cake to eat; the more the cake was spread out evenly the better, so the cattle didn't get bunched up, with the strongest ones getting the most.

If you knew where you were going to feed the stock, when the snow was deep, and if they hadn't spied you yet or smelled the cake, it was best to stomp down the snow before spreading out the cake so it wouldn't get buried in the snow, but there was seldom an opportunity. I tried to find areas where the wind had blown the snow off as much as possible. I'm sure the cows smelled the cake down in the snow and pawed every piece out anyway.

Branding the Calves

To be able to tell a rancher's cattle from anyone else's, all cattle carried the owner's brand as soon as they were old enough to be branded, which was well before weaning time. They also had an ear mark. Ours had the tip of one ear cut off, and then an inch slit down the middle of the cropped ear. There was one ranch that marked their cattle with a dewlap in addition to an earmark and brand. A piece of hide a ways below the throat was sliced down about four inches and left to hang, called a dewlap.

Our brand is a reversed 3 N, which stands for "the three nurses". There are so many brands in the state of Colorado that about all the easier brand combinations were already taken. It pays to keep your brand once it is owned by renewing it every five years with the state brand inspector. Our brand takes two branding irons: one a half circle to make the reversed three, and the other a bar to make the N. We had two of each, so that while one set was being used the other set was kept hot in the fire. The calves needing to be branded were kept near their mothers so they wouldn't become too alarmed, and they were driven into the small corral by the barn to be easier to catch in a crowded space. A pit eight to twelve inches deep was dug near the corral and a fire built in it. Chopped wood was stacked nearby to keep it burning hot, but the flames were kept low. When I was little, I sat on the corral fence during the branding and watched and ran errands. When I got bigger, I was allowed to tend the fire and carry the hot irons to hand over the corral fence to the person branding. When I became older, I roped the calves and held them down to be branded. Being the fire tender wasn't as pleasant as it may seem, as the fire nearly seared my face with the heat before the day was done. Roping, flanking, and holding down the calves was more exciting, but it had its pains, since I never got through a day without being kicked several times. Once I fainted from an especially painful kick to the back of my hand.

Cattle Stray Off

In the early years we only had fences between our place and the Sanford's to the east and the Red Top ranch to the north. The south and west remained unfenced, but usually when our cattle strayed from home, it was in the direction of the fences, where there were other herds of cattle that tempted them to find some hole in the fence. Cattle on those ranches also found ways to get into our pasture as well. A few frequently got in and mixed with our cows. The grass we did have was so vital to our own livestock that we couldn't afford to feed any of the neighboring livestock.

I was eight when my folks got Ginger, and I mostly rode her to get the milk cows in the first year. We knew every one of our cattle by sight and their names. Soon I was out cowboying, cutting out our neighbor's strays from our cattle and driving them off to their own land. I knew who they belonged to by the brand and the ear marks. Sanford cattle were always reliably branded + HH, but the Red Top bought some of their cattle in addition to having those born on their range, so their cows could have a number of different brands. The Red Top strays were taken out at a gate at the foot of Haystack Butte, or else one mile to the west at a gate with high gate posts. Either of their gates was very difficult for a child to open, being tall and stretched tight. I had to use a rope to pull the inside gate pole forward enough to get the top wire loop free. Closing the gates was even worse. I usually put the top back in the top wire loop first, as I was able to manage the bottom better. There were three gates to return Sanford cattle depending which gate was closest.

Strays didn't just walk easily out of our herd of cattle and drive nicely home; never such luck as that. When you found a stray or two, or even five or six, among our cattle, they had to be cut out from the group and driven a ways to one side. But they had no intention of going the way one planned, wanting to stay with our cows, so they'd use all their cow tactics to avoid being driven off. A single one would be scared to go alone, and tried desperately to out maneuver the horse. If there were two, it was even worse, as they'd split and run

Cutting out a stray.

in two directions, each planning to get around the horse quickly and return to the other cows. One time there were three head, each totally determined to go its own way and back to the herd. I decided that day that three was the most difficult number to cut. Once you got to four or more, they had a tendency to stick together. Cutting out strays was a difficult task, and it took a fast and smart horse to keep them together faster than they could split up and go in different directions. Ginger was a good, well-trained cow horse, and it was really she that taught me how to ride and how to turn a steer back quick. About all I could do was hang on to the saddle horn with both hands to keep from flying off.

Range cattle never leave their calves unattended. A cow generally grazes in sight of her calf. It's sweet to see a cow and her calf lying down together, usually touching back to back. Cows keep their calves well scrubbed with their rough tongue. They lick until the hair on the calf is glistening and laying in waves. After birth the calf is thoroughly licked, starting with the face first, probably by instinct to be sure it is breathing and nothing is stuck over its nostrils. With a herd of cattle, one cow is designated to be the babysitter, and may have a dozen or more calves in her charge, while the rest of the herd grazes some distance away. It used to be that ranchers did not de-horn their cattle and left them with horns for protection against coyotes (and even wolves in earlier years). Mother did see wolves during her first years of homesteading. In later years, when my parents bought their bulls, Mother got polled Herefords, which, without horns, do not have such a broad head, especially for easier calving of the young heifers.

Like the stray cattle that got into our herd, likewise ours at times got out and with other cattle. We counted our cattle every few days, and made sure we saw each one about twice a week. Since all our cattle were named, and we knew them all, even when our herd increased to a hundred, we knew which cow, or yearling, as was frequently the case, we would be looking for. If they couldn't be found at home among any of the groups of grazing cattle, then we assumed that they'd strayed off, and we started looking through the neighbor's cattle until the missing cow or cows were found. In all the years, and of all the strayed cattle, never once did we fail to find the culprit. Except for one time when one of Jeff's cows was running with ours. She had pinkeye and couldn't see well. I must have been twelve when this happened, and at the time was one hundred and twelve pounds. After hunting for her everywhere for three long days in the heat of early summer, I dropped to ninety-eight pounds, and never found a trace of that cow, high nor low, in arroyos, cedars, or prairie, nor in the years to come did anyone ever find her bones. All the others were always found dead or alive, though some took a great deal of searching to find.

I once found a yearling heifer at least twenty miles from home in the Red Top during the last week of October after having searched every grazing herd of cattle along the way. Having left early on a sunny day, I didn't think to take a coat. A cold wind came up late that afternoon before I could get the yearling home, and it chilled me to the bone. It was so cold that I was tempted to unsaddle, leave the saddle behind, and wrap up in the saddle blanket. Afterwards, I got the worst case of laryngitis and couldn't talk for days.

Heel Flies Scatter Our Cattle

In the spring, late March or April, heel flies would hatch out. They apparently bit cattle on their heels, causing them to kick up their heels, throw up their tail, and run for whatever they thought would rid them of these stinging pests. Evidently, close-growing cedar trees were the best shelter from these obnoxious flies. Nearly all of our cattle would completely disappear at this time. Our range was not fenced on the west and south, and this was the direction they'd take. Before I discovered their hideout, I would search and search and not find a single cow, and continued hunting farther and farther away. For many years they were found in the same place. They had crossed Cactus Flat and hidden in thick trees in a little canyon below the Hogback. There they'd be found tucked into the trees as best they could to keep away from the horrid heel flies.

The Longhorn Strays: A Girlhood Feat

Our land couldn't stand many strays, and for certain not a whole herd of longhorn steers. The rancher neighboring us to the southwest didn't have cattle of his own, but he made money from pasturing other people's cattle. At first he pastured Angus cattle, which hadn't been straying. He also pastured sheep, but had a herder. Stock could stray, as the range in that direction was still unfenced. One year he pastured a herd of Longhorns, and these critters were definitely the roving kind, with no calves to slow them down, long legs, lean bodies, and very long horns. Since the land was not their homeland, having just been brought in from out-of-state, they didn't have a homing attachment. It wasn't so much a desire for "greener grass on the other side of the fence" as it was the ranging instinct inherent in longhorns. It got to the point that a herd of sixty or more head of longhorns came down to our land every single night. Like a horde of locusts, they could eat the meager grass off our land in short order. That rancher didn't ride over and take them home. We had to, which meant me. Laddie was the horse with sufficient stamina to do it, since they didn't round up at all easily. He'd get the scattered, grazing bunches together and start them headed back up west. We'd get them moving down Highway 10, take them a couple miles down it, and then turn them south to cut across Cactus Flat, up to the other side, and from there south for another couple miles to near Averson Canyon, well into their own range, with the hope that they'd stay put after such a long drive and not come back, but do their nightly grazing to the west or south on their own land.

They didn't stay put at all, and early the next morning all would be back in about the same place they were the morning before. I got on Laddie as quickly as possible to round them up before they could snatch much more grass, and we would repeat the long drive all over again. I really doubt that any other horse would have had the endurance to get such a herd rounded up each day, single handed the way he could, and hassle all the darting steers that had plans on making a quick getaway.

Chickens

Having a small flock of chickens for eggs and meat was one of the vital priorities of starting over. When the neighbors had come to welcome my parents back home, Mother inquired about finding chickens. Two weeks later, Dad went and came back with three gunny sacks of chickens, a couple in each bag, one being a Rhode Island rooster. I no longer remember much about those first hens, but will never forget that rooster. He was nearly the size of a turkey tom and mean. When he saw me out in the barnyard, he'd come running to attack me and fly on my back as though I was a dangerous varmint. I would run crying for safety, and then have to endure Mother dabbing iodine on my back for the claw marks. Iodine was her medicine for every cut and wound. Dad said, "He never jumps me. Get a stick and stand up to him." I was only three, so it was easier said than done.

After this first start of chickens, Mother acquired a variety of chickens. The kind that was best for both laying and eating was the Rhode Island Red. Mother also wanted several hens that got broody frequently for hatching eggs, usually a large meat variety like Buff Orpingtons or speckled Plymouth Rocks, since we could use quite a number of fryers. After 21 days when baby chicks hatched Mother started them off with mashed hard boiled egg combined with a little raw oatmeal, and if any onions had started to sprout, she added finely cut green onion tops. We didn't have electricity and hence no refrigeration, so our meat was freshly killed. She usually only hatched with one hen at a time so all the fryers wouldn't get too big at once. If too many hens got broody at the same time, they had to be "broken up", as we called trying to stop a hen from being broody. Mother would put them in a gunny sack and hang them on the clothes line for a couple days. Young pullets were kept for layers for a few years. The older hens were usually boiled with homemade noodles, which I just loved, or at times she'd drop in dumplings. Mother never butchered a laying hen. She had a way to know by placing her fingers between their pelvic bones. If she could place only one finger, for sure they were not laying. If two fingers fit, they probably had recently quit laying or would likely begin to lay again. Three fingers was a sure sign of a layer. Hens that were laying also had a rosy comb and sang frequently. She put setting hens in bushel baskets with straw in a part of the barn away from the other chickens. When she began to raise turkeys, setting hens were used to hatch the turkey eggs.

Mother didn't waste anything and took good care of the hens. So among other things, she saved all the bones. She had a rock out by the chicken house where she'd squat down and hammer bones into small pieces for the hens to eat, since we didn't have oyster shell for their calcium. They must have craved the bone meal, as they all crowded around for the treat. One hen got her beak in the way at the wrong moment and part of her top beak got chopped off with the hammer. Mother felt sick about it, but the hen remained friendly and still loved chopped bones, although eating after that was a problem.

Many things were done for the chickens so they'd continue to lay during the winter. Keeping them warm was one thing. All the cardboard Mother collected during the year was saved for the beginning of winter and was tacked to the walls for insulation. Old rags were stuffed around the eaves and other places against cold drafts of air. In zero weather a few sheep were put in the chicken house at night to add their body heat. Their drinking water was never allowed to become icy, as Mother kept adding hot water from the teakettle. The chickens were kept in when there was snow on the ground so they wouldn't eat any snow. One extra cold, wet spring, I remember Mother keeping the setting hens in their baskets in our bedroom. One of them had been given duck eggs to hatch. One night, the first duckling hatched at midnight, waking us up, startled to hear loud, rapid quacking! In the spring, she often dissolved a piece of blue vitriol (copper sulfate) in their drinking water as a protection against internal and external parasites. The safe dosage is 1/1000.

Once Mother was checking on the chickens before she left the barn. Our "chicken house" was actually a big L-shaped room on the east end of the barn. There was a large rattlesnake inside, and it was still there when she came back with the shovel to kill it. We always killed rattlesnakes near the homestead, as their venom is deadly to humans and livestock. But more snake problems were caused by big, yard-long bull snakes, sometimes called gopher snakes. Mother wanted to keep them to catch the rats and mice in the barn, but she didn't want them to eat the chicken eggs. I remember watching, because she wouldn't let me go with her, as she drove a huge bull snake three-quarters of a mile away from the house, as she didn't want to kill it for eating eggs. Later that day, to her surprise, she found the same snake back in a chicken nest, and that time the snake lost its life. Another bull snake that lost its life was found with a long row of bumps inside it and a mother hen cackling nearby. Mother quickly killed it and cut it open in hopes that some chicks might still be alive, but all were dead.

The chicken house was cleaned regularly and the chicken manure brought to the garden. To prevent lice, Mother used crankcase oil on the roosts that the filling station in town gladly gave away. This was also good for keeping mites from getting in the scales of their feet. In later years she used to put a very thin line of Black Leaf 40 on the roosts on a warm night, the fumes of which were supposed to get rid of lice. In summer, mites were a problem. They'd be clustered together looking red from the blood they had sucked. For nests we sometimes used fifteen gallon oil drums given away by the gas station stuffed with soft dry grass. Mites got in these drums, and to thoroughly rid them of every mite, the drums were taken outside and a small fire built in each drum.

If singing is an indication of happiness, our chickens were happy. Hearing hens sing at the top of their lungs gave the ranch a pleasant atmosphere. Nor would a ranch seem like a ranch without the crow of the rooster. Mother once mentioned that chickens always seemed to lay more on Sunday. Even if the hens laid daily, I did notice more cackling of egg laying on Sundays. I know the day of the week has nothing to do with it, it just seemed to happen that way. Our flock could never have been "city hens", living in a little cubicle to lay eggs. They had the run of the barnyard and far beyond; they dug and scratched, and they ran chasing grasshoppers and insects for the majority of their food, plus any house scraps Mother tossed out to them. They wallowed in delectable dust baths or would spread a wing out over a leg as they basked in the sun. The hens kept their feathers nice by preening often. In winter, when there wasn't any dust, Mother often put out the ashes for the hens to wallow in, which likely helped get rid of lice too. The hens rewarded Mother for her care by giving some eggs all year long. I got to help gather the eggs, but if a hen was still on the nest, I was afraid she would peck my arm. It didn't seem to faze Mother.

Turkeys

Turkeys were another major part of starting over with livestock on the ranch. Turkeys had a threefold purpose: to help locate innumerable rattlesnakes, to save grass by cutting down the grasshopper population, and to provide additional ranch income. The start began with turkey eggs to hatch, which were set under the first three chicken hens that became broody. In later years, since when hens became broody was not dependable, Mother got a second-hand kerosene incubator that could hold about four dozen eggs. She put it in the cellar, where the temperature was more constant and there were no drafts. It was a regular battle to not lose the eggs, which needed to be kept around 100 degrees, as the thermostat had to be frequently adjusted and the kerosene flame would go out during the night, allowing the eggs to chill, unless it was discovered soon enough and reignited. I watched Mother turning each egg twice or oftener every day, like a mother turkey would turn them, for most of the twenty-eight days it takes turkey eggs to hatch. She also lightly sprinkled them with warm water for humidity. A turkey's under feathers get damp from the early morning grass when she's off the nest to eat, drink, and take a break, naturally giving humidity to her eggs.

The turkeys helped thin down the heavy population of rattlesnakes. Mother said she would keep a flock of turkeys around the ranch if for no other reason than to spot the rattlesnakes for the protection of her small daughter. And that they did. The first year that my folks had turkeys, they killed sixty rattlesnakes, forty or more the next year, and always around thirty each year after that. The majority were located by the turkeys, which would form a circle around the snake, coiled up ready to strike, and stick out their necks while emitting "gurr-oooo, gurr-rhuuu" sounds and gobbling by the toms. If it was a king, racer, or gopher snake, the snake would keep on traveling and the turkeys would follow alongside. My folks knew from the way the turkeys acted what kind of snake was found.

A second major reason was to protect what scant grass grew on the prairie during the drought years from hordes of grasshoppers. The grasshoppers, of course, also fed the turkeys—saving the expense of buying as much feed. Our young turkeys grew fast and became fat with such a high protein diet. Mother ranged the turkeys out, like one does a flock of sheep. To get their fill, early morning was best, while it was cool and the grasshoppers were sluggish. The turkeys would come home with bulging crops. It was necessary to herd them to protect them from ever lurking coyotes. Some mornings were hazy, even foggy, with coyotes ever on the watch to sneak in and grab a turkey from the edge furthermost from Mother. Turkeys chasing grasshoppers are hard to keep rounded up and become scattered out. Coyotes did manage to kill a few fat hens. Mother fought back with a vengeance, not wanting to afford any loss. The coyote menace occasioned one of those rare trips to town to buy strychnine from the drugstore. She put some in balls of fat and placed them inside the killed turkey carcasses. Later, I heard her say to Dad that she got some coyotes.

The turkeys were taken out again in the late afternoon, when the flock did a lot more chasing after the now active grasshoppers. The season also made a difference. When the grasshoppers were young, they were hoppers, but as they matured and developed wings, they gave the turkeys quite a chase.

Turkey hens are expert at hiding their nests. Chickens lay around the barnyard, mostly in the chicken house nests, but not so with turkeys—they ranged out around the arroyo and found secluded hiding places in the wild. Mother spent considerable time, me tagging at her heels, searching the arroyo hunting for turkey nests during the laying season. Once a nest was located, care was taken to not disturb the turkey by taking only some of the eggs when the turkey was gone, leaving the rest for the turkey so she wouldn't be suspicious and find a new, still better, hiding place. The eggs were kept cool and turned every day until a chicken became broody to set them under. Eight turkey eggs, being considerably larger than chicken eggs, were the most a chicken could adequately keep warm and covered. The turkey hens were set when they became broody and had laid a dozen or more eggs. Coyotes would catch them in the night if left where they nested outside the barnyard area. So Mother made wooden coops about three feet wide for them filled with dry grass for a nest. Next she had to catch the hen on its nest after dark at night and bring her, flapping and struggling, to lock up in the coop with only a couple eggs placed under her until she calmed down, and then the rest were placed under her. With a flock of small poults at her feet, a mother turkey would place each foot very carefully never stepping on any.

I loved to watch young poults, about a half pound in size, when they preened and dozed. They were learning to take good care of their new feathers. First they bent their neck around to extract a little oil from an oil spout on the top of the tail with their bill, then they spread it out on each individual wing or tail feather, starting with the base working out to the tip. They lifted a wing to oil the feathers underneath, or sharply bent their neck to preen chest feathers. If mother turkey stayed long enough, they'd close their eyes and doze, emitting a contended crooning. The little toms were really amusing once they got wing feathers, pretending they were a big tom turkey, as they strutted around a little demure female. Scratching an itch on top of their head made me laugh, standing on one foot, a couple claws of the other foot flew like lightning on the itch. Soon they were off again, as mother turkey continued teaching them to search for food.

In the sudden and hard summer thunderstorms, there wasn't always time to get everything under shelter before it struck. Little turkeys would get wet, chilled, and fall on the ground. They'd appear completely dead, but just to try to save them, even if they might be dead, they were brought into the house and wrapped up in towels, woolen rags, anything of warmth, and placed on the open oven door. It was amazing how they soon started to kick and every one came back to life as they got drier and warmed up. Evidently, they were only unconscious, but had not shown any sign of movement or breathing.

A pair of Great Horned owls that lived a mile away in the cave of the canyon would come down at night when the turkeys were a few months old and roosting outside. The owls silently landed on the branch the turkey was on and sidled up against it, without alarming it, for the kill. They were wasteful birds, for all they ate was the white breast meat. Mother began to sleep outside near the turkeys, if they couldn't drive them all into the barn at night. A few years later she built a shed to protect the turkeys at night.

One warm summer afternoon Mother walked out to the woodpile for an armful of wood to keep the stove burning hot for pears she was canning. She was utterly appalled, beside herself with worry, at the sight by the woodpile. There were her half-grown young turkeys all staggering and falling over, kicking and flopping on the ground. She couldn't imagine what disease had so suddenly struck these turkeys, and she just couldn't afford to lose them. Feeling sick at heart, her nurse's mind spun a mile a minute pondering every possible thing that could have so suddenly hit the entire flock. They had been fine yesterday; not one looked a bit sick earlier that day. Her agony lasted some while before she finally figured out what had happened. They were drunk, and most likely would all live. It must have been the pear peelings that had fermented before she tossed them out by the woodpile in the hot August sun, containing sufficient alcohol content to inebriate these greedy young birds, who had stuffed on the delicious pear peelings.

A third reason for raising turkeys was to earn money, a scarcity during those lean years. We raised Narragansetts, which are somewhat the color of a Plymouth Rock hen but more striped and not speckled like the chicken. They are handsome looking birds that are rare nowadays, as everyone wants the broad-breasted bronze or Beltsville white types. In the fall, when the turkeys were ready for market, Mother made arrangements with the neighbors to come help for a long day. With no refrigeration, butchering had to be done quickly. The turkeys were penned up two or three days in advance with no food, and the last day they also were not allowed any water. The turkeys were sold without being drawn (dressed), making it necessary for their entrails to be as clean as possible. Twine was hung from the rafters of the barn and everything cleared out in preparation for the big day.

Neighbors began to arrive before sunup. One of the men was the "sticker". He had a very slim, fairly long knife, and due to years of experience he expertly slipped the knife up the turkey's bill to poke an exact spot in the brain. This caused the turkey to "freeze" and its feathers to loosen. He then slit the turkey's throat and hooked a heavy weight in its lower bill to pull the neck down straight, allowing for good bleeding. Once the bleeding was nearly stopped, he loosened its legs from the rafter rope and hung it up for a lady to pick off the feathers. Then he caught the next turkey to be stuck and bled. There were six or more people removing all the feathers off the turkeys hanging from the rafters. Actually, feathers were rubbed off, leaving only the few pinfeathers to be picked out. Wing feathers are usually hard to pull out even with the turkey properly stuck. Mother received the finished turkeys and checked them over for any feathers or pinfeathers that may have been missed, and then wrapped the head in paper and placed the bird in the truck to be taken to market early the next morning.

I was small, about five years old. All day I managed to stay underfoot, wandering around in knee-deep feathers, looking up at the ladies flying hands as they rapidly removed the feathers before the body cooled and stiffened making feathers less easy to remove.

They were kind to me, saying nice things, and managed to not show annoyance. Curious of everything, I also closely observed the sticker, who probably would have liked to chase me away so as not to watch his job.

Well before dawn the next morning the old flatbed Chevy was loaded and driven off to Cheraw, the nearest meat packing house, one hundred mile away. Old white sheets covered the turkeys. On one trip Jeff was driving, and since he smoked, he threw his cigarette out the window. It landed on the sheets covering the turkeys and started a fire so that many of the birds were scorched before it was noticed, and they were unable to be sold, which was a sad setback when depending on every penny in those lean years. Several years later, Mother and Dad were taking the turkeys in our pickup, and this time somehow Dad was unable to see the road and they ran into the bar pit, rolled the pickup, and dumped out the turkeys. I remember Mother saying how all the dust and dirt on the floorboard of the cab went to the top when it rolled over and then fell down on their faces making it hard to breathe. The pickup had rolled completely over and back up on its wheels. They got the turkeys reloaded and continued taking them to market.

Mother hoped to clear one dollar a piece on the turkeys, and it was with this money that she purchased the land of the adjoining homesteads when the homesteader couldn't make it and had to sell their land. Turkeys were important to our lives, for in essence they are what added several sections of land to the ranch when no other money was to be had.

Times were difficult to start the ranch over from scratch and to get livestock and a garden established to feed ourselves and to make a living during those early years of the Great Depression. But the Depression, chiefly financial, was only one of the adversities we faced. During those same years the Great Drought struck, when very little rain fell, often in the form of severe hailstorms or blizzards in winter causing nearly as much damage as help. The grass barely survived due to lack of moisture, only to be eaten by prairie dogs with towns a mile in diameter or devoured by millions of grasshoppers.

The Grasshopper Army

One spring night in the early 1930s, I was witness with my parents to a phenomenon few people have experienced. It was around nine; we were in the kitchen part of Mother's two room, sod-roofed cabin; the lamp was turned low. Some heat remained in the wood cook stove. We were getting ready for bed when there were a few pelting sounds on the north wall, followed soon by thicker pelting. It was neither raining nor hailing. The wind was not blowing. What could this be? To add to the mystery, we heard a distant, eerie, very low-pitched rumble. My folks couldn't figure it out. With the lantern lit and jackets put on, they went out to investigate, leaving me in the safety of the cabin. Soon they were back, but the distraught, worried expressions on their faces gave a foreboding of something very wrong. They had seen something unimaginable. An army of millions of young grasshoppers was on the march, and a few of the front regiment had arrived. I'm sure my parents must have recalled the Bible stories of plagues, when hordes of locusts devoured the land. My folks slept very little that night.

At dawn we went out to examine what had befallen our land. The grasshopper army was still passing by; it was at least a quarter-mile wide and stretched back a mile or more. They were young, wingless grasshoppers, gray as the ground, and lined up in straight rows—row after row—each following less than an inch behind the next. The rows were spaced quite evenly two to three inches apart. The grasshoppers were all walking at a slow pace, no hopping out of line. Propelled by some unknown destiny, they were heading south. I stood near the west flank just looking, awestruck even at my age, not able to see to the other side or to the end of the troops. This gray-crawling mass of millions was starkly mesmerizing. It took a couple of days for the tail end of the army to pass on by, although many deserters and stragglers remained to populate our land. Where had they come from, how long had they been traveling, and where were they bound? Certainly no general could have better organized an army, yet this was all by some innate instinct of nature.

In the years that closely followed, termed the Drought Years, the ranchers were overwhelmed by grasshoppers devouring already scant grass. The government stepped in to help. Ranchers could get free grasshopper bait, probably just sawdust poisoned with arsenic. It came in big gunnysacks. My parents got several bags and scattered the poison over the prairie, particularly near draws where the grass was greener and the grasshoppers were thickest. But the grasshoppers were not all that died. After a rainstorm, runoff collected in potholes in the gullies where the cattle could drink for a day before the water sank away. The water contained enough arsenic that one of our beautiful cows died close to the water. Having been a nurse, Mother knew the signs of arsenic poisoning as she felt over the cow's prostrate body. Immediately, the rest of the cattle were moved away. Grasshopper poison was never again used on our land.

There were two places on the ranch of geologic interest for the possibility of oil. These areas had scattered geodes on the surface and nearly black soil from finely eroded shale. This soft earth was perfect for grasshoppers to sink down their ovipositors to lay their eggs a distance beneath the surface. One day, early in the fall, when I was out wandering around and checking on the cattle, I came upon a spot with thousands of grasshoppers matted together in twelve-to-eighteen-inch clumps depositing their eggs. These clumps of grasshoppers were spread out over a couple acres of ground. I dashed a mile to the house, breathlessly telling Mother about my amazing discovery. She quickly gathered piles of gunnysacks and wooden boards for paddles, and, with Dad, off we went. Intent on their egg-laying, these grasshoppers stayed put as we swatted them by the dozens. We loaded our truck with the bags of dead grasshoppers and took them home to feed the turkeys and chickens; the fresh, high-protein feed lasted for weeks.

Prairie Dog Threat

A typical animal of the ranch in the early days was the prairie dog. They lived in big colonies called prairie dog towns. Some towns were a mile or more across, but fortunately our ranch did not have towns quite that big. The first town I became aware of was in the east end pasture, and it was a good half mile across. These were doubly difficult years for my parents. The Great Depression was in full swing and they had little cash, and it was also a time of extended drought, prior to the Dust Bowl years. The grass was scant and needed for our flock of sheep, which they depended on for income from the lambs and the wool. The grass was also needed for our recently acquired, starving dairy cows. They couldn't afford hay or grain for winter supplement at that time. Although the prairie dogs were adorable little fellows and a symbol of the West, the other side of the coin was that they were grass eaters, and the land was too bare to sustain both the prairie dog town and the sheep and cows. Their colonies took up a large percentage of the land, so the prairie dogs had to go. The folks got gunny sacks of poisoned barley, and put a cup of this grain down every hole. It took a few years to rid the land of most of the prairie dogs. There was one small town in the south east corner of Aunt Henriette's section where they always remained, but the huge town that took up most of the east end was eradicated. Some remained for many years up near the buttes, and here and there a small colony of only a couple dozen holes could be found.

Dust Storms during the Dust Bowl Years

Often we'd have extremely strong winds with little dust. I could barely walk against them, and it was even kind of fun to lean forward and not fall down as the strength of wind would hold me up. The worst dust storms came during the Dust Bowl years. Mother said she could tell from which state the dust had come by its color. It was so thick that, no matter how tight the doors and windows might fit, the dust had a way of seeping in and coating the furniture with dust. The same thing was happening in other homesteads, and the ladies, to be slightly humorous, said the dust felt just like velvet. This amused Mother, who would say the same to me. It did feel like velvet.

Dusting did little good, except perhaps to make one feel like a good housekeeper. We dusted once a week, and when I got older that was mostly my chore.

When Dad took me to school in the mornings, the air was generally clear and the weather calm with a slight breeze. But by the afternoons, it was an entirely different story, when strong dust-laden winds blew in. Dad would come to pick me up from school before it became all but impossible to see through clouds of dust as dense as a heavy fog. It was a wonder how we got home. By now it was too dark to see the sun or in which direction it was located. I'd hang out of the passenger side of the truck and try to keep sight of the bar pit, which fortunately remained visible enough for me to guide Dad to hopefully keep him on the road. Today, roads have white or yellow lines painted on a their surfaces as guides, but these rough, pot-holed gravel roads had no paint. The bar pit is also called a "borrow pit", because in road building dirt was borrowed from the side to build up the surface, creating a ditch, where weeds, sunflowers, and tumbleweeds often grew from the moisture that ran off the road when it did rain. A few of my experiences getting home from school in the blinding dust were traumatic, as it was very slow going, and I had to try to keep what I thought was the road in sight and then not miss the turnoff to our home.

At home, the dust drifted behind the corral fences and any place that stopped the wind and caused it to drop its load. The barn could not be seen from the house; gates and doors were nearly impossible to open against the force of the wind; dust got in our eyes—we could barely see to do the chores and could only pray it blew over soon. One of the requirements to homestead had been to plow forty acres of the land and plant a dry-land crop. There was not enough rainfall to produce a crop, and irrigation water was non-existent. The fertile topsoil, formerly held in place by drought-resistant prairie grass, was exposed to the elements and turned to dust from plowing. Without enough rainfall to bind the soil to the ground, the dust just sailed away in any forceful wind that came along. These farming efforts were abandoned across the Midwest by the late 1930s, but it took many years for the grass to gradually reseed, to replace tumbleweeds and a brown weed that the stock didn't eat, and to regain a thick sod that the winds couldn't blow off. Mother prayed the wind would leave some of our soil behind. A hundred years later, 1917 to 2017, and our plowed forty acres have not yet completely sodded over.

Times were especially hard when my parents started over from scratch. As I look back on my mother's life, I see her as one of the most wonderful people I've ever known: truly made of "iron, stone, and beautiful flowers," as Aunt Henriette described her. Mother faced every trial without flinching, always figuring out ways to turn adversities into assets through her ingenuity and wisdom. Not only was this the height of the Great Depression, the great, decade-long drought was in full force, followed soon after by the Dust Bowl and its terrible dust storms. All of this drove many other ranchers away, but not Mother. Invariably, she stayed the course and kept us busy repairing, fixing, and making. Our hands would always be busy, our lives never boring.

Tumbleweed Children
By Letitia Black

I like to watch on a windy day
The tumbleweed children frolic and play.
Each daintily lifts its bouffant skirts
As whirling away, with the wind it flirts!
Racing each other without a care
They frighten the sheep and my skittish mare.
Their joyous antics, their infinite glee,
Enchant a blustery day for me.

Part III

Busy Hands

Pictured are block and tackle (we called it a wire stretcher) not only for fencing, but to pull pipe from the well or get a cow in a sling. We had several big pipe wrenches, a pitch fork (which was used to toss many a forkful of cut cactus onto the flat-bed Chevy for ensilage). My parents could fix everything with just an old hammer, pair of pliers, and baling wire.

Simultaneous with the process of starting over and getting livestock established on the ranch was the ongoing stream of repairing, fixing, nursing sick animals, making improvements, handling emergencies, and in general doing the chores of ranch and home. As it is said, "Never an idle moment."

Different sections of our ranch were named, which was helpful in describing where we were going or working. For example, Cactus Flat, the City Rocks, the canyon, the Woods pasture, the School section, the Mobly, the East End pasture, and the Buttes. Places where our livestock watered were the Rock Holes, the Seep, and the reservoirs: Jones, Bertha's, Barry, and Mobly. Mountains seen in the distance included Pike's Peak, Greenhorn, Old Baldy, the Spanish Peaks, and Fisher's Peak. We also named certain old cedars on the ranch after people (Will Black, Letitia, Jerry Simpson, Bill Henderson, and Barbara) to serve as landmarks.

Our Windmill

Nothing existed without water in this land of scant rainfall, so our lives and our livestock's lives depended on water drawn by our windmill. The well, which had been dug by Mother and Aunt Henriette, and its windmill were well along in years, but they are among those things that seem to outlast everyone. One necessary maintenance was packing the base of the ramrod with half-inch cotton rope, kept on hand for that purpose. As the old packing began to wear out, a new length of rope was cut, smeared with axle grease, and tamped in around the base of the ramrod with a screwdriver blade.

The iron pipes occasionally gave us problems. The worst was if something went wrong down in the well, a difficult and unpleasant place to fix something. Several thick planks were used to make a platform to cover the well. The planks were weighted down with a two hundred pound rock, which had to be maneuvered off the planks before entering the well. The ladder going down into the well was not very safe, as its rungs were always rotting from the constant dampness. The well was curbed with heavy sandstone slabs. Sometimes a pair of huge bull snakes would be in the rock curbing. No doubt to the snakes, the coolness of the rocks and the darkness of the well was an ideal place to retreat from a hot summer day. When encountered on land, I didn't mind them, but when climbing down the well close to them, they were frightening, especially when darting their black tongues out at me. The ladder ended on another plank platform about twenty feet down in the well. As a single length of pipe extended down the first section of the well, the trouble would be farther down where there were connections. We stood on this platform using two big pipe wrenches with 2½ - 3 foot long handles to unscrew the badly rusted 1½" pipes at a coupling. We had some lengths of pipe to fit over the handles, called cheater bars, to add leverage if necessary. Before the pipe was completely disconnected, a rope was secured around the lower section of pipe. Next we pulled the pipe from the well to find and fix the problem. The pipe was then tested to make sure everything worked properly and lowered back into the well to be screwed back together. The folks always managed to find and repair what was wrong. They had to, even if it took all day, or all week.

For the gears to run smoothly, one of my duties was to keep grease in the gearbox at the top of the windmill tower. This was always done on a day when the wind didn't blow, not even a breeze, so that the big windmill wheel wouldn't turn and smack me. Near the gearbox was the yearly nesting site of a pair of purple martins, whose fledglings assumed I was a parent bringing bugs for their gaping mouths. Climbing up the metal ladder carrying a few tools and the grease wasn't too bad. But coming down was more frightening. The ground always looked so far down. To keep my knees from shaking less, I would look up at the sky while descending.

Stock Tanks

A galvanized steel stock tank is very important on any ranch; it is the place the cattle and horses drink when water holes, ponds, or reservoirs are dry or not available. The tank is filled by the ceaseless up and down movement of the windmill's ramrod, which creaks and groans as the wind turns the big bladed wheel to bring water up from the bottom of the well. As a child, the water tank in our corral fascinated me, especially in the summer. All the varieties of water bugs that lived in water holes on the prairie were also found in our tank. I would stand captivated by the whirligig beetles, usually eight or more in a group; the water skaters, skimming on top of the water, agile as ice skaters; and the huge (to me at that age) oval-shaped brown beetles, skilled divers and swimmers. Growing below the water on the bottom were stands of nearly foot-high, fine-stemmed, non-leafy water plants, which provided hiding places for the insects as well as a couple of goldfish. I wanted so much to hold some of the beetles, but try as I would, they inevitably saw my hand approaching and off they darted to the other side of the tank. It was the same with the goldfish. When they surfaced to feed, it always was tempting to catch one, but they were far too clever and fast. For many years this was the only tank we had.

In order to keep the tanks from overflowing, Mother devised a float and plug system, similar to what our modern toilets have. Where the water came into the tank by an underground pipe from the windmill, she attached an empty, sealed gallon jug with a cord attached to a plug at the other end cut to the length she allowed the jug to rise and float until the pull of the cord would shut off the water to the tank.

The underground iron pipe continued uphill to and past the second tank, which we got a few years after returning, another thirty feet, toward an ancient cottonwood tree, connecting to the large storage tank, which never filled enough to overflow. From a gate valve at the bottom of the tank, water could flow by gravity back into the corral to the other tanks if needed, or to a long, low tank for the sheep to drink from. The storage tank provided a summer pleasure. If the folks had watermelon on a hot summer day, I loved to take my slice outside to eat, while thoroughly enjoying sitting at the base of the tank with my back against the always cool side.

We had water problems in the summer if the wind ceased to blow, which fortunately happened rarely. If the water holes and reservoirs went dry, water had to be pumped for all the livestock. The ramrod could be disconnected from the windmill and attached to a hand pitcher pump. Pumping the handle up and down at a steady pace brought up a little stream of water. To get enough water for the animals took tremendous and constant pumping, so the folks got a green gasoline-powered pump with wooden bars to attach to the ramrod when the wind didn't blow enough to keep water for the stock. It had a square

Cattle by the windmill.

pot in the middle to keep filled with water as a coolant. I was too small for a few years to help start it, but the folks always seemed to have trouble getting it running. In later years, when I was old enough to run it, I considered it the horrible, green, temperamental monster. But it must run, so that the cattle would not stand around bawling for water.

Winter was another thing. I don't know where the beetles went; the goldfish hibernated some place on the bottom of the tank. The ice would get very thick on top, so hard that Mother had to chop holes in the ice near the edges of the tank for the stock to drink. If a cold spell lasted long, with temperatures getting down to 26 degrees below zero, the ice might become six or more inches thick. At the second tank Mother devised a heater. The service station had given her some empty fifteen gallon steel grease and oil drums, about three feet tall. She would put a drum into the tank near an edge, weighted down with rocks so it didn't float, and build a wood fire inside the barrel. This helped to keep the tank from freezing over and warmed the water for the stock, as Mother did not want them drinking icy water when the weather was so cold. Neighboring ranchers chopped the ice so their cattle, already humped over with the cold, could drink.

The main galvanized water pipes were buried far down in the ground. To prevent exposed pipes and faucets from freezing in winter, Mother made wooden boxes to fit around them, stuffed the boxes tightly with thick layers of straw, then covered them to prevent them from getting wet and freezing. The exposed pipe from the windmill to the tank was wrapped thickly with burlap, although normally the water flowing from the windmill would keep it from freezing. Without running water in the barn or the house, there were no pipes there needing to be protected. In mid-December, when it gets dark and cold early, I remember Mother counting the days until they would gradually begin to lengthen.

Our Barn

The barn and barnyard were the hub of ranch activity, more so than the house. We practically lived around the barn when the livestock needed care. Our barn was already old when I was young. Its boards were aged and gray, its tin roof well rusted. I always wanted to climb on the barn roof. Mother told me to stay off of it because it was not good for the roof. Once I thought of a reason to go up. The boards by

the eaves were too rotten and they broke, causing me to fall to the ground on my back. The wind was completely knocked out of me.

Animals are hard on old barns, especially the cattle, as they are always knocking each other around and into the walls. Repair was an ongoing process, but a hand saw, hammer, pliers, and trusty baling wire performed miracles on many a wreck.

The front chamber of our barn with its main large door was comparable to a living room. A hay manger made from short cedar pickets at the north end was a favorite hiding place for hens to lay their eggs. In bad weather the cows were brought in there to be milked. It was a place to groom and work with the saddle horses. Many carpentry projects were done there, out of the hot summer sun or sheltered from freezing weather in the winter. It was just an all-year area of the barn whose support timbers were polished from years of cattle rubbing against them, with cow hair protruding from splinters. The kerosene lantern was carried to the barn after dark; I often held it to shine where Mother needed light. There were nails to hang the lantern up high, where its flickers of light made intriguing shadows on the walls. It was sheer joy to be in the barn with Mother and Dad, especially in the winter when dark comes early, surrounded by the soft sounds of resting animals, feeling snug and safe from the weather outside, bewitched by the warm dancing glow of our lantern.

Transportation

During the first five years after our return to Mother's homestead, the only roads in our area were two-track wagon roads, badly eroded in places, which often necessitated a new set of adjacent tracks to be formed. The folks first had the old flat-bed Chevy truck from California. Next, they acquired a black coupe, which they bought from Bill Henderson. I learned to drive in it along the fence lines, and Dad used to take me to school in it. It was cranky about starting in winter. For a while we had no vehicle until they bought a blue 1940 Chevrolet pickup, which lasted until I graduated from college in 1950, when Mother and cousin Jeffery arrived in a new blue 1950 Chevy pickup to bring me home. (I kept it over fifty years, and it was used to teach two younger generations to drive.)

We had little spare money to spend for gasoline, repairs, or nearly bald tires, so we usually walked or used a horse except when the car was necessary. My experience was chiefly with the 1940 Chevy. Our prairie had few spots where low-growing prickly pear cactus with long stiff thorns did not grow, making them difficult to avoid. The tree cactus did not grow so prolifically, and they could be dodged most of the time. We got so many flat tires from the prickly pear cactus that I quickly became a master at fixing and changing flat tires. On the day I got married I promised myself I would never change another tire! Sure enough, that afternoon on our way out of town a tire went flat. I stood in my dainty blue gingham "going away" dress as if helpless, watching my new husband at work, never offering any advice or help. Most likely I had changed a hundred times more tires than he had.

There was no such thing as taking a flat tire to a tire shop to be repaired. We just got to work to fix it: getting out the tire tools from under the seat, crawling underneath the vehicle to put the axle jack in place to jack the tire off the ground, wrenching the lug nuts off, and removing the tire. We used a leaf off the old springs to pry the tire off the rim so the inner tube could be pulled out. Some air was pumped into the tube to find where the puncture might be. If the hole was not obvious, the tube was taken to the water tank and held under water to watch for bubbles. When the leak was located, the spot was marked and the tube allowed to dry before scraping the surrounding area for glue to adhere better. We kept a little box of rubber patches to glue over inner tube holes. Before shoving the tube back into the tire, the outside of the tire was checked for a nail or rock, then we felt around the inside to be sure there were no thorns or anything else sticking out to puncture the tube once back in the tire. The tire was then put back on and pumped up. To be sure the tire pump always worked, we frequently oiled the leather gaskets so they didn't dry out. We never had an air gauge; we just "eyed" or kicked the tire for accuracy.

Until the later nineteen thirties, when Highway 10 was built, we had to drive on wagon roads that were not graveled and became muddy after rains or snow. In the winter, if a trip was necessary, my parents left at dawn while the ground was still frozen. Mother, who always tried to be prepared for emergencies, made sure they had chains, a shovel, old gunny sacks, and gloves along in case they got stuck, as there would be no one to pull them out. She even put in the lantern, in case it became dark before we got home. Chains didn't always prevent one from getting stuck. The ground would either be muddy or covered with snow when chaining up, making wet fingers stiffen and ache with cold.

Two neighbors used the wagon trail to town that commenced south of us and crossed Saunders Arroyo, where the bank slope was not too steep, just below our windmill and corrals, and drove through our barnyard on their way to town. The Nelsons had a car. Elderly John Flannery was more memorable. He went to town in his wagon pulled by Old Joe and Old Silver with Little Billy, the colt, trotting along, curious, yet never straying far from Old Silver, his mother. It would be at least three days before they returned, the wagon loaded with several months of supplies. Lou and Marie Nelson would stop for a brief visit on their way to town, but Mr. Flannery just drove his team straight on through our barnyard with only a grunted reply when we spoke to him. His cabin, some five miles to the south, was partially a dug-out with sandstone slab walls.

Town Day and Pin Money

The day and night before "town day" were very busy with much to do to get ready. All the empty gunnysacks in the barn were gathered and searched thoroughly for holes. Mother painstakingly darned every hole. The sacks were sold back to the feed store for a nickel each. The owner was always glad to buy sacks from Mother, as he knew there would be no hole left in hers where grain could spill through. The eggs were carefully cleaned and boxed, using a dab of vinegar on a cloth to clean them, as Mother said vinegar did not remove the natural protective coating on eggshells that preserves their freshness. The cream from our cows in five-gallon cans was so rich and thick it could be cut with a knife. We sold the eggs and the cream to the Fowler Creamery, and picked up empty replacement cream cans and egg cartons.

We had to be clean for town. The night before town day, Mother heated water on the stove and poured it in the galvanized tub for our baths. After doing chores the next morning, we had to have another quick wash. Clean, mended clothes were set out in readiness for morning. Mother had a dress she kept in the trunk and a pair of black shoes just for town. Dad and I each had just one pair of shoes, which were scrubbed and polished to look as new as possible to disguise their wear.

The next morning, after the stock was taken care of, we dressed. Mother had to put on a pair of ladies gloves to keep her work-worn hands from snagging her stockings. She put Tangee on her lips, light touch of rouge to her cheek bones, then face powdered her nose and face with a round powder puff. To me, her crowning touch was the pair of maroon beads she placed around her neck. Next she turned to me, striving for the impossible to get the last tangle out of my fine, wind-tossed hair. With my scalp still hurting, she gave my neck one last relentless scrub, then brushed my teeth with the toothbrush heaped with soda! If some dirt got on my face in town, out came her handkerchief, moistened with saliva, and the spot rubbed off.

Once we were ready and the car was loaded with our goods to sell, the car had to be started. This was before antifreeze existed, so the radiator was always drained after every trip to prevent it from freezing. To help the motor start easier, tea kettles of warm water were poured down the radiator. Mother always checked the battery to make sure it had sufficient water as well. For battery water, we used clean snow collected with a saucer, melted it in a crock (never allowing it to touch metal), and stored in glass jars, which we kept in the cellar. After Dad wore himself out cranking the old coupe, Mother took her turn. Sometimes in winter, when the motor was cold, it took a lot of cranking before it not only turned over but also kept running. Mother would sit in the car, foot ready on the gas pedal should the motor turn over, striving to keep it running until Dad could take over. My dad was not a swearing man; however, after an hour's ordeal of cranking, fingers numb with cold, and nose dripping, he would have a hot word or two to say to the silent vehicle.

At last we were off to town. In some places, the ruts were deep, and many arroyos had to be crossed on the way, especially where the road snaked around the same arroyo for a time, necessitating several crossings. Dad would gun the motor to go down the steep banks fast to get a "good run" to make it up the other side. This didn't seem

Crossing a steep arroyo.

to affect my folks, but my stomach didn't ascend as fast as the car, making it like a roller coaster ride. Even though the prairie was not fenced then as it is nowadays, there were close to a half dozen gates to open. Mother got out and opened the gates as Dad drove through; that is, if she could open them. Cowboys were tough, and one mark of strength must have been how tight and hard their gates were to open. It took both Dad and Mother together to open some gates. It was an unwritten cowboy code to always leave a gate like you found it.

The folks spent as little time as possible shopping in town, purchasing only the essential items Mother had written on a shopping list. It was necessary to get back to the ranch before dark, so no time was wasted. The coupe's headlights were dim at best, and there was always the possibility that they might go out. I don't recall Mother packing a lunch for town, and certainly she never considered eating at a restaurant. She did buy something at the grocery store to eat in the car, and one thing she never failed to get was over-ripe bananas. The last thing to do was getting gasoline at either the Sinclair or Conoco station, plus filling a five gallon can with gasoline to take home. The attendant always made sure the car had oil, and radiator and battery water, cleaned the windshield, and checked the tires.

When I was twelve Mother added my name to the checking account so that I could take the pickup to town on my own and do the shopping from her organized list. She trusted me to use the checkbook as cautiously as she did. There was never an issue in town of me not having a driver's license for another four years until I turned sixteen.

Mother: Nurse to Family, People and Livestock

As a young lady in Iowa, Mother had been a nurse and the superintendent of Mercy Hospital. But that was then, and now she was a rancher, who I remember as my mother. These are some of the incidents that required her talents.

Family Health

As for the family health, Mother was more into prevention than curing. Every winter she got a few bottles of cod liver oil, poured out a brimming tablespoon full each morning, and had me open my mouth. If she thought one of us might be getting a cold, we ate boiled onions for several days. She was careful with sugar, putting in half as much or less than a recipe called for, and definitely none was allowed in drinks (except homemade lemonade) or on cereals. Epson salt was kept on hand, not just for a laxative, but for soaking feet and other needs. A box of chocolate ex-lax would come as an advertisement in the mail once in a while. It tasted like candy, so Mother was careful to keep it out of sight. Iodine was her favorite medicine for wounds and punctures along with stinging rubbing alcohol.

Mother always had castor oil on hand, and if needed a teaspoon was given as a laxative. I hated its taste. Sometimes I got a stye on my eyelid, which she punctured with a needle to press out the pus and then put on a dab of castor oil. It also worked well to relieve chapped skin. She used witch hazel to soothe irritated skin, too, and always had a jar of Vaseline and a small tin of ironite available for sores.

For getting something out of an eye, she held up the eye lid and told her patient to look down to the corner of the nose. When it was really hard to locate, she rolled the eye lid up on a match stick, which I never liked, to find the obstacle. She kept an eye wash in a sterilized jar made from boric acid powder dissolved in boiling water. Mother used olive oil to clean our ears of wax, making q-tips from a tooth pick with a bit of cotton wrapped very tightly around the end. She was very careful to not shove anything down the ear opening. When a letter arrived from a friend who had a sickness in the family, we weren't to touch the letter, and Mother scrubbed her hands while grumbling about people sending germs through the mail. If someone came by whom Mother suspected of having a venereal disease, door handles and jams, out-house surfaces, and everything they might have touched was scrubbed before I was allowed to touch things. I remember thinking sunshine was healing, and if I was feeling sick I would go out by the barn to soak up the sunshine.

Blood Poisoning

One year when we had trouble in the bottom of the well, we used Manila rope to pull up lengths of pipe to fix. I must have gotten a rope sliver in my palm, probably bruising as well from the weight of the heavy pipe, causing what we called "blood poisoning" in my left palm. A swollen lump formed with radiating, red streaks up my arm. Unable to sleep, I paced the floor during the night, even fainting a couple times from the stabbing pain. Mother did her best to pull the infection and poison out of my hand by applying hot poultices of stale bread soaked in milk and some honey with other things that pulled as they dried, maybe raw potato or Epson salts. She would apply this warm, damp mixture between layers of cloth to not only pull out the poisons but to aid in circulation of the area for healing by alternating heat and cold. It took time, but eventually my hand healed.

Les Valdez

In later years, Jeff had land south of ours and ran a band of sheep. We had added our flock to his so that we no longer had to herd them. He and my parents had hired a sheepherder named Les Valdez to herd them, and he lived in a covered sheep wagon that could be moved as the sheep were pastured in different areas of the range. Les was a large, tall, lanky Aztec-Mexican who always ambled with a limp. He had a few bristly whiskers on his chin, and the only teeth I noticed were two front fangs. They did look just like fangs and protruded from his mouth. I have no idea how old he was, but his hair was fairly gray.

One day around noon we saw Les limping hurriedly toward the house. He was in great pain because he had not been able to urinate. The tighter his bladder became, the worse he hurt. Mother immediately got the wood stove going with small pinyon sticks for quick heat, put on a kettle of water for juniper berry tea, and a big pan of water to warm for the tub. With pants off, she had Les sit in our big, galvanized tub in warm water, while she poured water from a pitcher held up high to give the sound of running water. The thing I recall most was Les's bony knees sticking out of the tub and his moaning. In less than twenty minutes Mother's combined three treatments gave him relief. Les, grateful and smiling, got his pants back on, sincerely thanked Mother, and headed back to the sheep.

Tenth Little Pig

When I was about five the folks acquired a huge white pregnant sow. Two weeks later, she gave birth to a squirming litter of ten piglets, mostly white like herself. A few had patches of tan. The ten were a great fascination to watch when mother sow, content for all the world, lay down flat on her side, to expose a belly lined with teats, to suckle her brood. All ten squealed, grunted, and vigorously shoved each other around with feet and snouts trying to get the best, fuller teat of milk. At least she had plenty of teats to go around. By the third day, Mother observed one tyke all bloated up. Although Mother picked it up very quietly to not arouse the sow, it instantly began letting out shrill squeals. Sows are known for being one of the fiercest of ranch animals when their babies are in trouble. Fearing the corral would not hold up to the fury of this giant sow, Mother quickly ran carrying the pig across the barnyard and on behind the house, where she hoped the squeals would not be heard at the barn.

Having been a nurse, Mother was a capable veterinarian in animal emergencies. She looked the pig over and discovered that it had no bowel opening below its tiny curled-up tail. Everything it had eaten remained inside. While Dad held the pig, undaunted Mother got out her nurse's scalpel, boiled it for sterilization, then sat on the back doorstep with the piggy across her lap. With painstaking care, she determined just where to make an incision to be exactly over the bowel opening. This tenth little pig of the litter never let up squealing, even before the cut was made. Dad was standing by, and I was torn between the scene of the operation and nervously peaking around the corner of the house every minute to see if the sow got out and was coming to get us.

I remember how pleased Mother was when her incision on the pig's soft, little, pink behind was true to the mark; it was exactly over where the little tube-like bowel existed under the previously smooth flesh. The tenth little pig squealed and squirmed like it was being murdered, but it must have felt relieved when the accumulated feces began oozing out. It was brought back to join its litter mates, and it grew up to become a fat little pig, just like the other nine.

Rattlesnake Bites

At times some of our animals would get struck by rattlesnakes. We could always tell because the area around the bite would be tightly swollen and the animal lethargic. Mother was successful in saving them at the time, but later in winter some died, apparently from having their heart weakened. For leg bites she often soaked the area with either turpentine or kerosene. On a dog bitten in the face, she used bacon grease and honey smeared on a cloth soaked in turpentine. Poor dog, its whole head and neck were so swollen that it struggled to get its breath. On big lambs she usually gave antivenom shots. Soon they would fall off asleep. Mother wasn't sure if they were sleeping or dying at first and felt anxious for them until they woke up. She said in case she should ever need to give antivenom to a person, it was helpful knowing what to expect by having given it first to animals.

Cold and Wet

Calves by the oven door.

Mother could usually recuperate cold, wet chicks and turkey poults that appeared to be completely dead. Surprisingly, once they regained some body heat, they would start to twitch and move. She did the same for freezing lambs, and even once a calf. After they were carried into the house and laid on a gunnysack by the open oven door to warm, they usually lived.

Cow Sling

Once Mother drug posts and materials out on the prairie a mile from the house to build a sling for a cow that had gotten sick and was too weak to stand. With a block and tackle (our fence stretchers), she had Dad help get the cow to her feet and in the sling. Being on their feet seems to give animals more interest in life. She carried water and corn fodder out to the cow several times a day until she got it well again. Another time, she put a sick cow in a sling in the barn and nursed it for a month until it could walk again.

Sick cow in a sling.

Sheep Tallow Salve

One extra hot, dry summer even our dog's feet cracked and bled. Mother prepared a salve made of sheep tallow (rendered in the oven from fat trimmings when sheep were butchered) and honey to rub on his paws. This was the same salve we used on our cracked lips. She also made him booties out of heavy canvas and tied them on his feet to keep him from injuring his feet worse.

Scorched Flour and Boiled Skim Milk Cure

Sometimes young calves would get the scours, which is a yellow-gray colored, smelly diarrhea. Her cure was to scorch flour in a skillet, making it similar to charcoal, mix it in boiled skim milk, put it in a bottle when cooled, and pour it down their throats; not an easy feat with a struggling calf. Calves with scours always recuperated a day or two after Mother's treatment.

Animal Midwife

Many are the nights Mother sat up with a sick animal. Sometimes a calf would be born in breach position. The cow, knowing it was in trouble, would allow Mother to reach her arm deep up inside her uterus to maneuver the calf around to a position it could be born. She never had to resort to having to "pull" a calf with chain and block and tackle. Oftentimes she also had to help an ewe struggling to give birth. One time Mother failed was when our brood mare, Babbie, died having her last colt. Poor Babbie tried so hard to give birth. Mother went to assist her where she was giving birth in the draw above the arroyo. Babbie, expecting her help, kept turning her rear end to Mother begging, but Mother was unable to remove her colt. It was too late, and both the mare and the foal died. Mother always grieved for Babbie.

Building Fence with Mother

The ranch had boundary fences to the east between us and the Sanford land, and to the north between our land and the Red Top ranch. There was a fenced horse pasture, but most of the rest of the land was open prairie. During my growing-up years, Mother and I built many miles of barbed wire fences for both pastures and the boundary. It is likely we acquired the heavy, three foot tall rolls of used barbed wire and the three hundred plus cedar posts for every mile of fence we built through Bill Henderson, who must have brought it to us from the Sanford ranch, as they had removed many miles of unwanted fencing from abandoned homesteads. We also had taken down about a mile of unwanted fence and rolled up the barbed wire, weaving it in and out so that no wire loops slipped off the roll, to use elsewhere. How I helped Mother depended on my age at the time. At first, I just held posts straight while she sighted the line and filled the post hole, tamping the soil down firm with the crowbar head. Later, I also did everything, except I never sighted the line when Mother was present. Her fences were straight as a die for miles on end!

Building fence with Mother.

A tape measure was never used; we didn't have one. The distance between the posts was "stepped off", usually sixteen feet (about a rod, which was common usage back then), sometimes twelve feet or less if the terrain was rocky or uneven. A shirt pocket or some part of the body was used to determine the height of the top wire. The distance between the fence wires was measured with the hammer. Most cattle fences were three strand. Sheep fences had to be four strand, with the bottom wire being much lower than for cattle. First, we laid out the posts where they were to go, and then we dug the holes eighteen inches deep, putting the dirt in a tidy pile nearby. Post holes were easy to dig on the ranch, as the soil was not unduly hard and there were seldom any rocks. We used only a shovel and a crowbar to loosen and dig the dirt.

At the end of each fence line or wherever there was to be a gate, we put in sturdy, end cedar posts, which were taller and larger in diameter than the line posts and inserted about twice as deep in the ground. For a brace in the fence line, we used what is called a "dead man." This was a huge, squarish rock with several strands of barbed wire wrapped tightly around it that was placed in a pit about six feet past the bottom of the post. From the rock, a double wire was attached securely to a notch about three-fourths of the way up the post. After the rock was buried and tamped tightly in the ground, the two wires were twisted with a stick in the middle until they became taunt. Then another brace was put in the fence line facing in the opposite direction of the "dead man". This was just a heavy, straight post angled from the ground to near the top of the brace post. A strong corner post that couldn't give, Mother would say, was the key to a long lasting fence. For gates, an H-brace was usually used.

The next step was to roll out the strands of barbed wire along the post line. We rolled the strands out several feet apart so they'd not get tangled. Tough gloves were needed to roll the wire out carefully, weaving the roll in and out a bit so no big loops of wire would slip off while rolling it. At the first brace post (usually less than a quarter mile) the wire stretcher was attached to the top wire. We started with the top wire so that in case there was any "give" in the brace post, it wouldn't affect the lower strands. There was generally so much slack at first that a new "bite" had to be taken a time or two before getting the right tightness. Generally, we fenced in warmer weather, so it was important not to get the wire too tight; otherwise, when the wire contracted in winter, it could break. This never happened to us, but cold weather often popped staples out.

Until all three wires were in place, four if a sheep fence, we put in a few staples, partially driven in, to hold the barbed wire up in place until we were sure of the even tension of all the strands. Then we went back and stapled the wires behind a barb to the post, measuring their distance apart with the hammer handle.

Gate posts were not only large around like the corner and brace posts but needed to be tall so that gates could be recognized from a distance. (Nothing like having to ride a mile or more just to locate an inconspicuous gate some place you'd never been before!) If the posts were tall enough, there would be a double strand of barbed wire twisted between the very tops of the gate posts as a brace. The gate itself would just be barbed wire strands attached to two light-weight posts with wire stays every couple feet to keep the wires separated. One end of the gate was attached to one of the big gate posts, and the gate's opening end post was set into wire loops at the top and bottom of the other main gate post to keep it closed. These loops were sized so that it was always a struggle to close the gate; they had to be tight enough so some cow couldn't get it opened. On some ranches where there was no family, just a cowboy, I thought they tried to make their gates tight, but what had probably happened is that the gate posts had pulled back due to insufficient bracing, making them harder and harder to open. I always carried my lariat for such gates, using it as a pulley.

It was inevitable that at times fences crossed over arroyos. When the summer cloudbursts occurred, the arroyos would flood, causing a very strong torrent of water that took out everything in its path. So to win in any way at all, water gaps were made to cross the arroyos.

These were essentially just a long gate with light wire stays every three to four feet apart, weakly attached with bailing wire at both ends that could break with the force of the flood. In such a case, the gate-like water gap would be thrown off to the side. After a flood, the water gaps were generally a mangled mess, but at least yards of fence were not destroyed on either side of the banks. Once, after a huge flood in an area where the land was shallow on both sides of the arroyo, a quarter mile of fence was utterly destroyed.

Using old wire required a lot of splicing. Our method was to bend back the end of one wire about eight inches, making a small loop, and then twist it back around the wire. The end of the spliced piece was stuck through the small loop and also twisted tightly back around its wire. In washes or gullies, we hung sizable rocks from the bottom wire, and wove a wire stay from the top fence wire down to keep the lower wires from sagging and evenly spaced.

Fixing Fence with Dad

The ranch property was in sections. The outside fences followed the section lines, so they were at least a mile long, and two miles if two sections were in sequence. Dad liked to take me with him for company when he checked the fence line and to hand him the staples as needed. He also wanted me to find those that had fallen on the ground and were hidden someplace in the grass. In those years, we did not waste anything; even a crooked staple was of value once it was hammered straight.

About the time I was eight the folks had an old, black, used coupe that Dad drove on the prairie when he fixed fences. He'd start at the farthest end and work toward home. To save having to walk the mile or so back to get the car, he'd have me drive very slowly after him by propping me up on the seat and setting the throttle to just creep along, as my legs didn't reach the floorboard. Mostly, I just steered around all the tree cactus and yucca clusters and wasn't to get into any holes or washes. I knew how to turn the ignition off to stop the vehicle if necessary. Dad appreciated all the walking it saved him, but I felt very proud to get to drive the car all by myself.

WPA Men Build Highway 10

Among the ways President Franklin Delano Roosevelt created jobs during the Depression years when so many people were out of work was to establish work programs like the Public Works Administration (PWA), Works Progress Administration (WPA), and Civilian

Conservation Corps (CCC). The government paid men to work on projects that would benefit the country, one being road construction. In our part of the state, there was no road to any town from our ranch and the other ranches in the area, only two track wagon roads that wound across the prairie. In the mid 1930s, under the WPA, the construction of Highway 10 was commenced, creating a direct route from Walsenburg to La Junta, Colorado, about a seventy mile stretch where no roads existed.

As our ranch was about midway between these towns, my parents were asked if they would board and house from twenty to thirty men during the week days while approximately twenty miles of the highway was being constructed in either direction from our house. There is a joke about workers leaning on their shovel. It's true, as I saw plenty WPA men leaning on their shovel handles when I rode a horse out to watch them at work.

There was also a crew of CCC men, who were mostly young and unmarried, but they did not board with us, and they worked independently of the WPA men. Their part in the highway construction was to build the bridges. They quarried the limestone outcroppings along the Sanford Hills. Their masons cut huge rectangular blocks of limestone, which were hauled to the bridge sites to make the side walls. Their excellent work still stands today, strong as ever.

Needing a way to make some money, the folks agreed to take on boarders. The crew foreman had men with some carpentry skills build a bunkhouse attached to the west side of our house with two-tier bunks to sleep thirty men who brought their own bedrolls. Since we had no bathroom facilities, the men used our outhouse. With them using our toilet, Mother did not want us to use it, fearing diseases, but Dad still did. Mother and I went down out of sight to the arroyo. The men washed up by dipping water from a bucket into tin wash basins. A dining hall was built in the right angle between our two-room homestead cabin and Aunt Bertha's big one-room homestead house (which had been moved and attached to the back west corner of Mother's cabin after Bertha married and relocated). A flagstone floor was laid with big, flat rocks hauled from where the CCC men were quarrying limestone. Mother was an expert at maneuvering and placing various shaped flagstones to fit together nicely. The room was walled, roofed, and screened as simply as possible, and down the center, a long plank table was made with benches on either side to seat thirty men. She bought three dozen plain white plates, cups, and cereal bowls, along with pitchers, big serving bowls, platters, and silverware since we had few dishes. It was also necessary for her to purchase very large kettles and skillets to cook food for so many men.

What Mother accomplished was miraculous—cooking for such a big crew and still keeping up the ranch. She had only one wood cook stove, no electricity, no running water, and a small kitchen not equipped with big cupboards, counters, or drawers like we're accustomed to nowadays. Over the kitchen table, which was her main work space, she made a rack to hang the utensils she used most, so they could be grabbed quickly—every movement had to count! Her timing and organization were perfect. On the table she rolled out pie crusts, kneaded loaves of bread, peeled hundreds of potatoes, sliced the meat, and prepared everything. Pinto beans were a mainstay of the men's home diet, which Mother cooked in huge kettles. One of my chores was to carefully sort the beans on the back side of the table. Beans came in one hundred pound burlap bags and were not cleaned as well as now. There would be bean-sized rocks, small clumps of dirt, and even tackbur stickers among the beans.

At meal time, I set the plates and silverware around the table, and Dad carried out the bowls and platters of food and did the general serving, though I was tall enough to keep their water glasses filled. The men, Mexicans, mostly family men from the coal mining town of Walsenburg, served themselves by passing the bowls of food down the table to each other while Dad saw to it that the bowls were kept full. I remember one man who was large and had a huge belly who always ate three times more than any of the others. If any of the crew did not show up and there was room at the table, I was allowed to eat with the men, which made me feel proud. I was attracted to one of the older men, so I'd check to see if there was a space to sit next to him.

Evidently these men didn't have a lot to eat at home, and definitely nothing like the food Mother served. Mondays they ate huge quantities and tapered off as the week progressed. Mother baked twelve loaves of bread every day. Each night the bread starter was put in two big crocks to raise overnight to be ready for the next day's baking. She alternated between making cakes or pies for dessert—all baked in our one small wood-fired oven in quantities for thirty hungry men! Even when we were not dairying, a few cows were always milked. There was fresh butter, pitchers of thick cream (thicker than store-bought whipping cream), and pitchers of milk at each end and in the middle of the table.

By four o'clock Mother was up to have a hearty breakfast ready. There was the steaming coffee pot that held two gallons, slices of bread, bacon and eggs some days, other times pancakes or cereal. Before the men left for the day, thirty lunches were waiting to be taken, also prepared by Mother. By mid afternoon she would be starting the evening meal and be up late at night preparing food for the following day.

There were heaps of dishes to be washed. I well remember washing dishes. After the meals were cooked, a copper boiler was put on the cook stove to heat water. The dishpan was our big galvanized tub, set on a bench. After Mother had warm water and soap in the tub, I washed the dishes, putting them in a dishpan to rinse, pouring hot water over them from the tea kettle, then moving the clean dishes to another big milk pan to drain, drying them each time it got full. On school mornings, I usually didn't have time to wash breakfast dishes or do milking and other ranch chores, but I always washed the evening dishes.

There was one problem that Mother intensely disliked. The men were to pay $1.00 a day for their room and board at the end of each month, after they got paid. But a few did not bring their money, causing trips to Walsenburg to go around to their homes to collect from their wives. This embarrassed Mother in addition to being a terrible, time-consuming inconvenience. Often a wife did not have the money to pay until a future trip to town. I remember their small adobe-type homes and how cleanly swept their hard packed dirt yards were kept, usually with small children out playing. Another problem Mother had was with the men wanting to bring me, this only child, treats of candy, which she tried to confiscate and hide without my knowing. Illness or dentists couldn't be afforded; it was necessary to keep us healthy.

Another difficulty that annoyed my clean Frisian mother to the extreme was that apparently some of the men brought bedbugs in their bedrolls. Even though the bunkhouse was at the opposite end from our bedroom, somehow bedbugs got into our beds. Mother fought them by searching every crack and cranny, taking the bedding and mattresses outside to sun, and even used poisons, but it was many months before we saw the last of them. They were definitely difficult to eradicate. Not long after the men had finished this section of the highway and left, the bunkhouse was taken down and the lumber put in the board pile behind the barn. Even without the men, it was likely to continue to harbor the dreaded bedbugs, so it was decided to dismantle it. I recall doing most of the dismantling of the bunkhouse with pry bars and a hammer.

The Grass Fire

One hot summer day shortly after noon, Mother and I were surprised to have our sheepherder, Les Valdez, come rushing in. He was so out of breath we were unable to make out what he was urgently trying to tell us. At last we made out that he had accidentally started a prairie grass fire when trying to burn out a rattlesnake near his camp wagon. He knew the danger of a dry grass fire, and had left the sheep to hurry as fast as he could five miles for help. Except for a limp and shuffle of one foot, he got around all right using a cedar walking stick, which to a sheep herder is as useful as a lasso is to a cowboy.

As luck would have it, Dad had taken the pickup to town that day. Mother and I were alone and ten miles from the nearest phone. I quickly got and saddled our only horse, Ginger, while Mother gathered up every shovel and hoe and a number of gunnysacks. In moments, we were on our way to the fire. Les was given the horse. At his age with his lame leg, the rough trot of the mare must have been as much an ordeal for him as his rush on foot to get help had been. Mother and I alternately ran and walked fast in spite of the blazing summer sun. Although she was then fifty, mother still could run very well.

Arriving at the scene of the fire, Mother could tell it was already spreading too fast for the three of us to control. She ordered me to ride Ginger on from where we were to the nearest neighbor's place, six miles farther across-country, and then down a canyon to phone for help to all the neighbors that might be home. I was to contact Jeffery's dad in town. As soon as she finished giving directions to find the Nelsons' place, Ginger and I dashed off. It was far too hot a day to gallop Ginger all the way, but I hurried her as fast as I dared. In spite of a few hard-to-open and harder-yet-to-close fence gates, and the uncertainty of which canyon fork to take, we made it. Luckily the Nelson's were home. My biggest worry all the way over had been to find them gone and have to hunt for another neighbor's home with the distance and time involved, although mother had prepared me with what to do. Mrs. Nelson phoned all the neighbors that were at home. I had never used or talked on a phone and was terrified with the thought of having to do so if the Nelsons had been gone and their house open. I would have had to try to phone with no one to show me how. Mrs. Nelson gave me a cool drink, before hurrying back to fight the fire.

All this time Mother and Les had been desperately fighting the fire to keep it from spreading, never stopping for a moment. Once back, I began to toss a shovel full of dirt on higher flames and beat lower ones with the shovel head. One could only make a few licks of progress before jumping back to keep one's face from being seared by the heat.

By late afternoon, help started to arrive. My throat got a lump in it seeing the neighbors come jolting over the sod in their pickups, snaking around the bigger cactus and yucca clumps. Stopping some ways back here and there along the fire line, each jumped out and started beating the creeping flames with no words spoken. By their constant fight Mother and Les had kept the fire from spreading more than a mile wide, and with the help of the neighbors it was put out before dark. Luck had also been with us that day, for no strong winds blew.

The neighbors were thanked, and one of them took Mother and the tools home. Les stayed and gathered up the sheep, which hadn't strayed too far from the area. I rode Ginger home in the awesomely colored twilight caused by the smoke in the air. It certainly had been an eventful day: my first fire, riding to find the Nelson's place, and the scare of having to use a telephone. It was typical of neighbors that unfailingly you could count on them to come give a hand when they were needed in times of hardship, drought, dust storms, illness, or catastrophe.

Early the next morning Dad, Mother, and I went over in the pickup to see if Les was all right and to be sure the fire was dead. Les was mumbling about "rheumo". He had "rheumo" real bad. I guess rheumatism was his way of describing aching joints and sore muscles. No doubt the horseback ride had contributed to some of his aches and pains as well as the firefighting. Later that day Jeff came out from town to check on what had happened, and Les said, "I guess you fire me now." Jeff replied, "It's too late now."

Laundry

Modern day laundry is so simple, so easy. Clothes are put into the washer, a few buttons punched, and the clothes are washed, rinsed, and spun, ready to dry. In the dryer, a few more buttons pushed, and soon the clothes are dry, ready to fold and put away. Count your blessings when you compare it to how we did laundry when I was young.

Mother was careful when washing woolens that they never got in hot water to cause them to shrink; she used only slightly warm water. Two galvanized tubs were put on a bench on laundry day. The copper boiler was filled with water and heated on the wood stove. Hot water was poured in the first tub, containing the wash board with a bar of P & G soap on its top ledge. The other tub had lukewarm rinse water. If the piece to be washed had bad soiled spots, extra soap was rubbed into the stubborn areas. Clothes were washed by rubbing them up and down on the corrugated surface of the wash board as they were dipped in and out of the sudsy water. Then they had to be wrung tightly by hand and put in the rinse tub. Not having another tub, we used a dishpan for a second rinse. Rinsed clothes got an extra tight wringing twist before putting them in a bucket to carry to the clothesline.

We couldn't wash whenever we wished. Washing was timed to occur after a good snow or rain, when rainwater could be collected off the roof or snow scooped from a drift. When rain started, we scurried around to get all the buckets, tubs, and barrel under the eaves to catch the drips. The first few minutes of runoff from the roof was generally dumped to get rid of dirt and debris. We didn't use the well water to wash clothes because it was so hard it caused the soap to curdle like sour milk. Waiting for rain in that country often meant a long time between wash days. By then most of our bedding, clothes, and towels were in the laundry. If possible we waited for a sunny day to get everything dry before evening. Our clothesline had four heavy wires spaced two feet apart on 2 x 6" boards, stretched and bolted to two huge cedar posts placed thirty feet apart. In order to make room for everything on the clothesline, all the light towels and linens were hung up first in hopes they'd be dry before the heavier things were ready for the lines. On hot breezy days the first things would dry in a jiffy. The only other dryer we had was a small line stretched below the ceiling in the kitchen, as a last resort in winter.

To keep linens white or for stains, they were boiled in sudsy water prior to washing. The main way Mother bleached, especially in summer with the hot sun bearing down, was to spread dish towels and other whites out on top of the alfalfa. The intense sun bleached as well as sterilized. Under the clothesline she had planted a small patch of alfalfa for a number of purposes besides that of wash day. If clothes fell off the line, especially if a sudden gust of wind whipped through the yard, or were dropped while being hung up, at least they didn't land in the dust. In addition, the alfalfa served as our lawn and was the only green feed, except for garden weeds, we had to give the chickens so they'd lay eggs with deep yellow yolks. After the alfalfa had blossomed and was mature to cut, it was dried for a treat should an animal get hurt or sick.

Mother made her own soap by saving all the fats left over from cooking. With boiled mutton, she set the broth aside for the tallow to form a thick layer on top which could be skimmed off. Fat hens cooked for chicken and noodles produced lots of yellow fat. Besides lard, the hogs that were butchered rendered more fat. Before long, Mother had a number of five pound cans of salt free grease saved up to make soap. Her mother, and generations before her, made their own soap. No recipe was needed. A can of lye was dissolved in cold water in the big crock using a wooden paddle. I watched, but I can't remember what else was added prior to slowly adding the liquefied grease and stirring it to the desired consistency before pouring it into trays to cool. When set, it was cut into bars. Mother always liked her soap better than the store's P & G bars.

After six or more years, one fall when the lambs were sold, my folks decided they had enough money to buy a washing machine. Why, you might wonder, did my mother go through the drudgery of hand washing for six years? Credit cards or monthly payments were not an option back then, nor was getting a loan against something for collateral feasible. My parents never spent a dime they did not have. They had no regular monthly income to make budgeting easier. Money only came with the seasonal sale of lambs or calves, and it had to last throughout the year, allowing for ranch expenses of hay, grain, and everything else.

Mr. Morgan's store sold Maytags with gasoline motors. This made wash day far easier for Mother. But nothing is perfect because this machine had a way of not wanting to start the morning of wash day! We'd be in a hurry to get an early start with many loads to wash, and wanting to get clothes on the lines quickly so everything would dry by evening. We mixed gasoline and oil according to the directions, turned the levers just right, then gave a good quick pull on the rope to start the motor, always with full anticipation that it would start within the first half dozen pulls, or if not, that it would at least give a little kick to show some kind of life. When nothing happened, we speedily tried remedies from the easiest and most likely first to the more complicated. Fortunately, the easy remedy—the spark plug—was often the solution to the trouble. The carbon had to be carefully scraped off with the blade of a pocket knife. It was essential to have the gap correct, which was the width of a dime. In the worst of cases, we lost half a day working on that stubborn motor, but once it gave in, it ran fine the rest of the day. We did have to be careful using the machine's wringer so that not too much went through at once. Buttons had to be positioned so they didn't get caught, and fingers and hands had to be guarded from the rollers.

Mother was very saving with our precious rain or snow melt water. The soapy water from the washing was used for as many things as possible. First, the dainty things were washed, next white things, then the more soiled, ending with the darker colors, and lastly our blue jeans. The water was then carried to the plants. The next round of water was usually that of the first rinse water along with new rinse waters. When the last loads were done, the rinse water was used to mop the floors before it was given to plants. Mother loved her Maytag and carefully cleaned it after each use.

Our linens were starched and ironed. Mother put several tablespoons of flour in a large kettle, added a bit of cold water, and stirred until the mixture was smooth. Then she added more water to boil, stirring continually until the starch was quite transparent. Often she had me stirring while she did other things. When done, cool water was added. After the last rinse, clothing or linens to be starched were dipped into the starch water. Those requiring the stiffest starching, such as the curtains, were dipped into the starch water first, then the rest. All were tightly wrung out for the clothesline. If she could catch the clothes to be ironed at the right moment, she took them off the line while they were just slightly damp and rolled them in a towel for ironing. Otherwise, a few hours before being ready to iron, the clothes were sprinkled and rolled in a towel.

Mother had three flatirons (also called sad irons) of different weights, which were heated on the wood stove. A hot fire was made to heat all three irons until they hissed with a drop of water. When ironing, she dampened a finger tip on the end of her tongue and touched an iron quick as lightning to judge its temperature. A piece of wax paper was kept on the ironing board to wipe the iron clean. She ironed until the iron was too cool, placed it back on the stove, and took the next one, rotating the iron she was using with the ones heating. She knew by the hiss of the iron how hot it was, so as to have the hottest for linens and a cooler iron for woolens. When not in use, these three irons always stood on end at the way back corner of the cook stove. When older, I ironed following Mother's example.

One quality about snow water, when it is used for laundry, is its complete softness. It took all the dirt out of even seemingly clean clothes. There were city children a couple years older than I whose mothers boxed their outgrown clothes and sent them for me to grow into. The clothes were always laundered before being sent, but if we washed them again in snow water, the water would have a sooty tinge and the clothes came out looking even cleaner. Amazed, I asked Mother how come. She said the city smoke in the air accounted for it. Although rainwater is soft, our rainwater had the color of very weak tea, likely tinted by tar on the roof; it was not as absolutely soft as the snow water.

House Cleaning

Having few possessions, our daily housework was accomplished quickly, at least with Mother's swift movements. Our dust cloth was lightly oiled and kept in an earthenware vase. She liked using turkey wing dusters too, and an ample supply was collected each fall when the turkeys were butchered, keeping the tip-end joint of the wing with its long wing feathers. Our lamp and lantern chimneys were always washed first, before the dishes, in the sudsy dish water to keep them sparkly clean for better lighting, as they frequently got sooty streaks if the wicks were not kept very even. The slightest peak would give a higher flame and consequently make black soot in the chimney. The wicks were trimmed when needed and the lamp and lantern refilled with kerosene when low, always to be in readiness. Another thing Mother kept shining clean were our two windows with small panes, one on the north and the other on the south side of the kitchen. She used vinegar and shined them with newspaper. The putty that held the window panes in place often cracked from the harsh weather and had to be replaced. She rubbed linseed oil on it to make it last longer.

(She also used linseed oil to rub on shovel, hoe, and ax handles to prevent them from cracking and getting splinters.)

I used to stay up some nights to thoroughly clean the cupboard and all the shelves while Mother complained that I should be in bed. With the wood stove, the walls got dingy. Mother used kalsomine (calcimine), an inexpensive green tinted wash, to paint them every year. Once she bought an ivy printed wallpaper for the kitchen walls. I remember she cooked a paste (flour boiled in a small amount of water until translucent) to glue it on, but the wallpaper peeled back down in places before it had dried enough to stick. She had to redo those places and hold the wallpaper in place until it dried. The wallpaper livened up the kitchen and we liked it.

The pantry shelves were kept tidy with an ever watchful eye for any sign of mice droppings. She also checked the closet shelves, where all pillow cases and towels were neatly folded and put away after wash day. Mother's "nice" things were kept in the trunk, such as embroidered napkins and pillow cases, which were only used when we had company. She had a lovely blue and white patterned quilt her friend, Barbara Padgett, once made for her. It, too, was kept safely in the trunk and only brought out as the bedspread for company. Besides mice, the closet had to be carefully watched for clothes moths. When not in use, the heavy woolen quilts were kept in the closet. She dusted them with black pepper to deter moths, and she also hung all the woolens and bedding on the clothesline on very hot summer days to kill moth eggs or larvae. Once a week, when the weather was nice, our bedding was aired on the clothesline, and once a month Mother turned the mattresses. When our felt hats needed cleaning, she rubbed them with dry cornmeal, and if not too beat up with use, they looked as good as new. (She said in the hospital if a patient was too ill to have their hair shampooed, their hair was cleaned with dry cornmeal.)

The cistern for our drinking water was cleaned once a year, or sooner if it got empty. It had to be ready for the first good snows and next year's drinking water. As soon as I was old enough to be dependable, I helped to clean the cistern. As I was small, it was easier for me to slip through the small opening. A ladder, used only for the cistern, was put down with a rope. I climbed down with boots scrubbed clean, a bucket, rags, and scrub brushes. Mother bailed up dirty water and sent down clean water to me as needed. To keep the cistern water sanitary, the bucket and rope that bailed the water were kept in the pantry in a pan covered by a towel. When bailing up water for drinking and cooking, one had to toss the bucket down just right to dip it full of water and not land on top of the water to float. A big metal nut was attached to the bail of the bucket as a weight to help it tip.

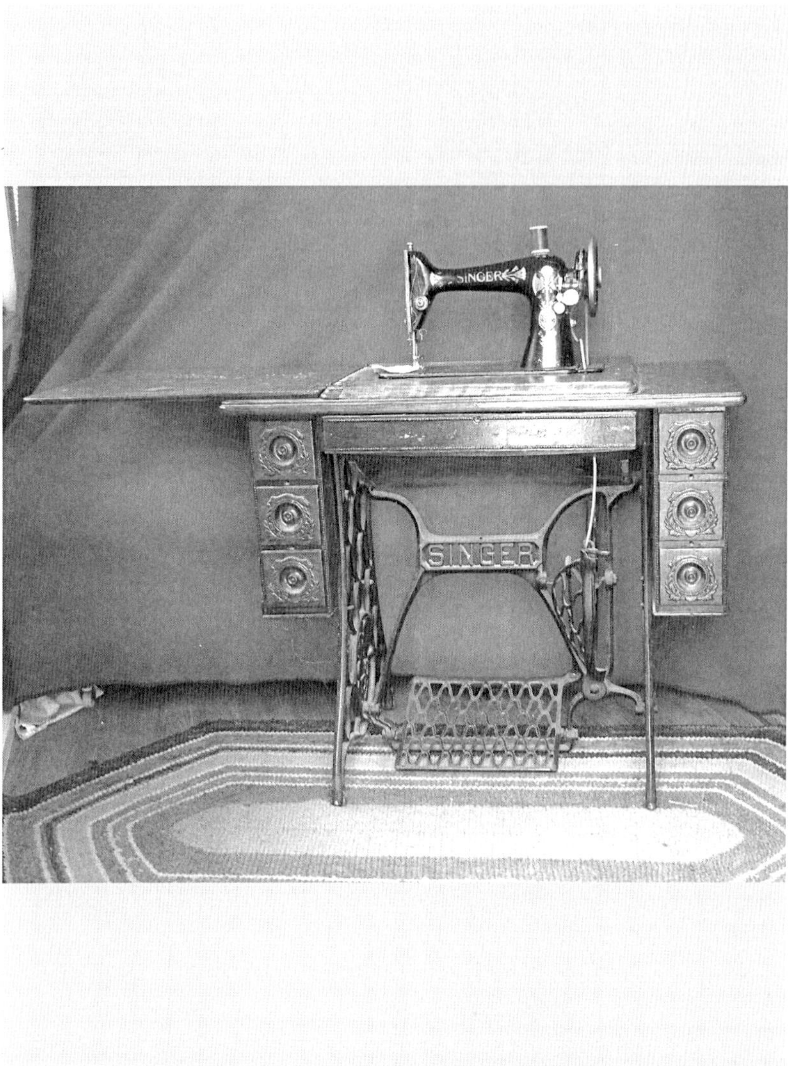

Dad had brought the Idaho treadle Singer sewing machine from Idaho (he had purchased it about 1894) when he came to Colorado. Mother used it constantly with all the mending ranch work clothes needed. She made new dresses and aprons, and managed time to make pillows, mattresses, and quilts. Under the sewing machine is one of the braided rugs.

PART IV

Homemade, Forever Frugal and Saving

When Mother made us new dresses or play suits for me, she used pieces of the left-over material to make the 'flowers' for this quilt. Nowadays it's a pleasure to reminisce over the clothes we once wore. Sandy Blake, our rural mail carrier, will finish the quilt for me in 2018

Being Frugal

Mother was extremely thrifty, a necessity of the times, which I took for granted as the natural way of living. Perhaps this quality was also inherent in her genes from frugal Frisian ancestors. With so much other ranch work to be done, one might wonder how she had time for baking and preparing well-balanced meals. For one thing, she had all the recipes in her head and made adjustments with what ingredients were available as she went. She did not use a recipe book nor a variety of kitchen gadgets, just a one and two cup measure and a regular teaspoon and tablespoon, if not a pinch or the cup of her hand for salt, pepper, spices, and leavening. All were swiftly mixed and sifted into one bowl and quickly ready for cooking or baking, leaving scarcely any dishes to wash.

No food was wasted. I remember how thinly she peeled potatoes to not waste any. At mealtime, we served what we could eat, second helpings if we wanted more, always leaving a clean plate. To conserve on fuel and time, Mother generally cooked large batches of food. Leftovers, such as pinto beans, were served again, often several days in a row, or combined into a new dish. Any scraps of food that could not be eaten were fed to the chickens or put in the garden for fertilizer. We never bought wrapping paper for gifts, but very carefully opened presents to not damage the wrapping and then ironed out any wrinkles to reuse the paper, hoping no one would notice.

Time was a precious commodity. Mother told me never to waste time, setting an impeccable example. The kitchen was arranged so no motions were wasted or time lost. No matter how many things required doing in a day, by organization and some miracle everything was finished by various time-saving formulas and juggling of tasks. She tried to arrange work so frequently two things could be accomplished simultaneously.

Money was carefully spent. There was very little in the first place, so it had to be used sparingly and wisely only for necessities. For example, we never took a trip just for pleasure or spent money for entertainments. We lacked home comforts others would consider necessities, and never ate in a restaurant or bought prepared food from the grocery store. Mother made most of my clothes and patched them when needed on the sewing machine. To save on car expenses, the folks took only necessary trips to town, maybe once a month when a list accumulated. The principle was, if you could do without something, then you didn't buy it. I can hear Dad saying as if it were yesterday, "We may not have much, but we don't have any debts."

Water was a necessity every homestead lady expertly learned ways to save; Mother was no exception. She practiced this art to the 'T'. Soft snow water was hard work to acquire. To collect clean snow, we waited until the second or third snow of the season after the air was very clean, as early snows usually contained dust. Clean, deep snow drifts were chosen, but not those close to the corrals or the house, which necessitated carrying the snow for some distance. Back and forth my parents trudged, each with two buckets of snow. Most bucketfuls were dumped directly into the cistern, but some were put into the boiler on the stove to melt. The warm snow water was poured into the cistern to help melt the snow. It took many days, and usually two or more snows, to fill the cistern to the brim with deliciously pure snow water, a quantity needed to last for our drinking and cooking until the following winter. To prevent contamination, all the buckets and shovels used to collect the snow were thoroughly cleaned and scalded.

The folks hoped for rains to fill the rain barrels for baths, laundry, and other household needs. When rains didn't come, we often had to use the cistern water. We knew well not to be wasteful with it, as it had to last the whole year. A dipper full was enough to wash one's face and hands. Afterwards it was emptied into the slop bucket and carried to water plants. Our Saturday night bathing routine made use of the water to the fullest. The water was heated on the stove and poured into a galvanized tub. I was always given the first bath, next Mother used the tub, and Dad's bath came last. We each had our own bath towel, like the three bears: small, medium, and large-sized. Mother used our bath water to hand wash a few clothes, then mopped the kitchen floor with it. She always mopped with a rag on her hands and knees. (She said it was good for the abdomen.) The last of the bath water was given back to the land to water thirsty plants. The outhouse didn't require water, and laundry was saved until rain came to fill the rain barrels.

One manner of saving that I never used, one I heard Mother mention frequently, was saving your energy. I had energy galore and loved to run more than anything. It was fun to speed right at Mother, stopping just before bumping into her, to be a tease. She'd say, "Some day you'll wear your heart out." Many years later I understand better what she meant, feeling as she had felt, tired from a full day of hard ranch work. She practiced ways to conserve energy whenever possible.

Like water and food, old clothing was saved too. When clothes could no longer be practically patched, the better pieces, and especially scraps left from sewing, were used for quilt making. Poorer material was cut into inch-wide strips, the ends sewn together, and rolled into large balls to be crocheted with a big wooden hook into our everyday rugs. Buttons, snaps, and zippers were carefully removed for future use on other garments. The very functional "rag box" had numerous ranch and home uses. When old rags got dirty, they were washed and used some more. Some discarded garments found their way back into use for patching holes in our clothes. String was always saved when opening grain sacks and rolled in a ball. String often came in handy.

We used a kerosene lamp and lantern. There were no flashlights. My parents used the lamp and the lantern sparingly, since our five-gallon can of kerosene needed to last. The lantern was carried when livestock needed to be cared for after dark. The kitchen felt cozy and friendly when the lamp was lit. On winter nights the wick was turned low unless a brighter light was needed for reading. Electricity did

Doing school work by lamp light.

not arrive in our area until I was in college. All my early schoolwork was done at the kitchen table by the light of our lamp. It was very important to be saving with matches. Some were stored in a metal, mouse-proof box.

The original homesteaders had to be a hardy lot to survive the unsparing elements nature continually flung at them. To remain in this harsh land, it took every manner of saving and ingenuity. Of the dozens of homesteads that once dotted the prairie as neighbors to my mother, only four have endured to be passed down to posterity. Mother's homestead is one of them. Unlike some of the others, who had income from family who worked outside, during this time Mother and Dad managed to make a living and support the ranch strictly by working on their land.

Mother's Aprons

During my younger years, Mother generally wore dresses she had made herself, almost all from the same favorite pattern. It was not customary for women to wear pants back then. Mother made the bib aprons, which she always wore, with two sizable pockets. Besides keeping dresses clean, they came in handy in many ways. In the house, it was quicker to grab the bottom of her apron than a potholder to hold something hot, or use it to wipe away a tiny spill, sweat off her forehead, or worst of all to wipe dirt off my face. From the woodpile, she'd bring in an apron full of kindling to start the fire, from the chicken house an apron full of eggs, from the garden an apron full of vegetables, or from the cellar an apron full of potatoes or onions. She carried little chicks or turkeys around in the pockets of her apron, or used the pockets to bring grain to some animal. The pockets would have things like her handkerchief, maybe seeds for the garden, or nails for a carpentry project.

Cooking and Meals

Naturally, cooking is associated with my mother, as outdoor ranch people have ravenous appetites. Mother managed her old, wood-burning cook stove with expertise. The weather could affect how the stove worked. Sometimes the wind gave it a good "draw" (draft), and at other times the humidity kept the smoke from rising and the stove barely burned. Once a traveling salesman came by claiming to fix stoves, and he insisted that our stove was in bad need. Mother didn't trust him, but she decided the fastest and safest way to send him on his way was to allow him to "fix" our stove.

Mother could whip up a meal of fried chicken in a jiffy. When she saw someone coming for a meal, she quickly put the tea kettle of water on to boil, went out and caught a young rooster with a long, stiff wire hook, chopped its head off, dipped it in scalding water (but not so hot as to tear the skin) to loosen its feathers for plucking, lit a piece of newspaper to give it a singe to remove hairs, cut it in pieces, and had it ready for the frying pan all in short order. She served roast turkey with plenty of everything else at Thanksgiving when relatives came. (Thanksgiving was the one time we had cranberries. For a relish Mother ground up a pound of cranberries with 3 oranges, 6 large apples, and 2 cups of sugar. For a sauce she dissolved 1 cup of sugar in 2 cups of water, added a pound of whole cranberries, and cooked for ten minutes over medium heat.) Her wood cook stove dinners were always the best I ever tasted. Neighboring cowboys often rode in for a meal, a welcome change from their own batching.

If something needed to heat up quickly, she'd remove one of the small round stove lids, burn small pinyon sticks, and place the kettle over the direct flames. When baking in the oven, she knew how to concentrate the heat there just right by stoking the fire and adjusting the dampers to get either a hot or moderate temperature. The oven had no thermometer, so she tested the heat by quickly putting her hand inside. Wondrous things emerged from her oven with an aroma that made one's mouth water. How often I waited impatiently, saying "Isn't it done yet?" for bread, raisin buns, cinnamon rolls, oatmeal cookies, chocolate cake, or pies to come out. Her sunberry and apple pies were famous. I often took pieces of apple pie outside to eat down in the arroyo against a shady bank. Aside from emptying the stove's ashes, around the oven had to be cleaned of soot periodically using a metal scraper. There were lidded openings in different places, especially behind the oven, to reach in and pull the soot out. (This is the black soot she used to mix with linseed oil to make paint for our sheep's brands.) The cook stove was our source of heat for warming the kitchen, the room we lived in most of the time. On a cold, windy day when getting chilled, Mother would say, "Nothing feels better than a good wood fire."

We placed great value and importance in matches and made sure there always was a good supply, which were kept in a mouse-proof tin box. Mother would say, "we'd be blowed up without matches." Nowadays cigarette lighters are available, but we depended solely on matches to light the fire, lamp, and lantern. She was careful to use only one match to start a fire or light the lamp or lantern. When she made tiny fires to warm us while herding sheep in the snow and cold, she took care that her tinder would catch on the first try, having the next size pieces ready, and was alert to cup her hand to ward off a breeze from blowing out the match or lighted tinder. She carried a small bottle with a few matches in her pocket when we went out.

Mother's cooking utensils were few: some baking pans, her set of Wear-ever aluminum kettles, a colander, and a couple iron skillets. Her skillets were "non-stick" from long usage. Homestead women never allowed their skillets to get in the dishwater; instead they always wiped them clean. Not having a rolling pin, a round bottle was used. She saved any batter left in a bowl using a thin metal spatula about an inch wide. It had been used so much that it was extremely thin and flexible, making it perfect to wipe a mixing bowl completely clean. I recall her saying that she could bake one or more cookies from the dough other ladies left sticking in their mixing bowls. On rare occasions when she left a little extra batter so I could lick the bowl, I often got hiccoughs. She said flour was a raw starch, not digestible until baked.

Our breakfasts were always hearty, mostly bacon, eggs, pancakes, French toast, oatmeal and raisins, usually slices of Mother's bread, and sometimes biscuits. Occasionally, we had a special treat of stewed prunes topped with our thick cream. We never bought dry cereal. Mother said there was more food value in the whole wheat we got for our livestock for only pennies than could be found in Kellogg type cereals. She soaked whole wheat in a thermos of hot water overnight to soften, then cooked it the next morning and served it with cream. In place of cereals like Cream of Wheat, she ground wheat in the coffee mill and cooked it as a hot breakfast cereal.

Supper was when we were through working outside for the day, and it was quite simple, nourishing and tasty. She made rice and raisins by cooking raisins in the rice and serving it with a dash of cinnamon, a sprinkle of brown sugar, and milk. Potato soup was a couple slices of bacon rendered in the bottom of the kettle to give flavor, then diced onions and potatoes were added with a small amount of water. Milk was added as soon as it was cooked, and the soup was eaten with crackers. For a bread and milk supper, bread was toasted in the oven, then buttered and covered with scalded milk; for a special treat brown sugar and cinnamon were sprinkled over the toast before the hot milk was added. Sometimes supper was cornbread; other times a skillet of fried potatoes with slices of raw onion; or hominy with milk; maybe bean soup; or a real favorite, chicken soup. Chicken soup was made from the bony rib and back parts, simmered a long time. When dressing chickens, their legs and feet were kept and scalded to remove the scaly covering, then boiled to make a healthy gelatin to add to chicken soup along with chopped carrots, onions, and rice (celery, too, if we had some).

Mother explained that we should eat seafood once a week for iodine that our bodies needed. She purchased canned oysters and clams to make into soup or chowder for supper. Canned sardines and salmon were served cold at the dinner meals. I loved salmon and the delicate bones in it.

Our main meal, called dinner, was in the middle of the day and generally included meat. In winter, when the weather became cold, a beef or mutton was hung up, and Mother cut steaks off the carcass and fried them in the skillet. They were so delicious! Our cool weather meat was frequently roast lamb, or lamb stew. When butchering

sheep, hogs, or a calf, a singletree from an old wagon was hooked into their hocks to spread their hind legs. A block and tackle hung from a beam was attached to hoist the animal to a convenient working height. My favorite dinner meal was fried chicken from the young fryers. I liked the drum stick and thigh best. When Mother dressed an older hen that was no longer laying eggs, it was boiled and served either with dumplings or her homemade noodles, which consisted of flour, an egg, and the water measured in an egg shell. She'd knead the stiff dough, then roll it out very thin on the floured table and leave it to dry. Then she cut the noodles in very slim, long strips and dropped them in the boiling chicken broth. One time the Morton Salt Company millionaire visited our ranch, and he complimented Mother heartily, saying how he loved her fricasseed chicken (which was the old red rooster and noodles).

I was observant of Mother when she butchered chickens, and so learned a lot about the inner anatomy of the body, since their organs were similar to our human insides. Some of the older hens were very fat, the fat even coating their intestines, liver, everything. Their livers were not strong and firm, but instead pale and fragile. Way back then as a small child, I swore to myself that I'd never let my body get fat like those old hens. She always saved the heart, liver, and gizzard, which when opened to clean often contained fascinating polished stones, and once even a small purplish dinosaur gastrolith.

Fried rabbit was scrumptious. From the frying pan drippings, with a spoonful of flour stirred in and milk added to the proper consistency, she made rabbit gravy, which we put on a slice of bread or mashed potatoes...so yummy. There were cottontail rabbits around the barn, board piles, and the arroyo. I used to spend hours figuring out how to outwit, corner, and catch one. It was always with much pride that I brought a cottontail to the house for Mother to fry for dinner. Those meals were absolutely wonderful.

Mother made sure we had vegetables cooked and raw with meals. She tried to have both a green and a yellow vegetable, saying they had different vitamins and minerals. She considered beets and carrots good for our fingernails and hair. Greens were healthy, as she noted that deer ate greens and were agile, flexible, and didn't have trouble giving birth. She planted carrots, beets, and salsify in the garden, which were kept in the ground over winter covered with straw and dug up a shovel-full at a time as needed. She dried her sweet corn to make fresh cornmeal. In winter, our cooked vegetables often came from jars of vegetables she had canned. When a jar was opened, she first checked to make sure the jar was still sealed, next smelling the contents to make certain it smelled safe. Tomatoes or fruits could be eaten cold from the jar, but she boiled all vegetables rapidly for ten minutes to prevent any possibility of ptomaine poisoning. Never once through many years and eating hundreds of jars of canning did we get poisoned. Without refrigeration, she also didn't take chances with leftovers becoming spoiled and boiled them too for ten minutes. Mother was clean, but she did not scrub vegetables to death to get off every speck of garden dirt, saying there were minerals in it our bodies could use. We never took vitamin or mineral supplements, depending only on our food to stay healthy.

When we had watermelon, Mother saved the rind and cut off the white part to make watermelon preserves with cloves and other spices. It was one of my favorite treats, and some of the preserves went in a delectable fruit cake. Her chocolate, sour cream cake was amazing as well as her wonderful upside-down cake with a layer of prunes, pineapple, and brown sugar on the bottom. Aside from her favored apple pie, she made mince meat, raisin, sunberry, and lemon and cream pies. Older eggs were used to make the meringue for lemon and cream pies. She said they whipped faster and stiffer. Oatmeal cookies were a favorite, but she also made sugar cookies that puffed up, ginger snaps and gingerbread men, and Santa Claus cookies at Christmas from an old Frisian recipe. Sometimes Mother used a skillet of hot oil and made corn fritters, potato chips, or doughnuts: I got to eat the round doughnut holes. I can taste them now.

Homemade Dairy Products

When the folks didn't sell cream any more, and milked only a few cows, the milk was strained into big milk pans for the cream to rise to the top to be skimmed off. Today people stay away from heavy cream, but we just enjoyed it. In winter, Mother made ice cream by freezing it in a snow bank. In summer, it was made in an ice cream maker, usually on the Fourth of July, when we had company. Relatives brought a bag of rock salt and a big block of ice, which was smashed up by the butt end of the ax and packed around the ice-cream canister, which was cranked until the ice cream mixture was too firm to crank anymore. The paddle was taken out, the ice cream scrapped off, and my cousin Jeffery and I got to eat all that was left on the paddle. Then the ice cream maker was set in our galvanized tub so the melting saltwater would not kill any plant life. No ice cream has ever tasted as wonderful as the real homemade ice cream of my childhood. The recipe wasn't fancy and was measured by guess: just the heavy cream, some beaten eggs and milk, with vanilla and sugar added. We frequently enjoyed having custard, rice pudding, and bread pudding with raisins for dessert.

The flavor of home churned butter is excellent. In the beginning Mother had only a half gallon jar to churn the cream. She churned it by shaking the jar back and forth over her lap while she sat to read. Every so often she opened the lid a crack to let the air or gas out. Soon the butter fat stuck in little clumps, separating from the buttermilk. The buttermilk was poured off. We all loved buttermilk. The butter was put in a bowl and pressed (we called it "worked") with the back of a spoon in cool water until it was firm and all the whey pressed out. The butter was then salted and ready to use. Fresh homemade butter tastes superior to store butter, comparable to the flavor of a tree ripened fruit to one picked green to ripen in the store. We later got a churn with a handle for turning the paddle. When a cow is first fresh with milk, her cream turns to butter quickly and it has a deep yellow color, especially in the spring when there is plenty of green grass. Later, when a cow is about to dry up, it seems like one has to churn and churn before the cream finally turns to butter.

Cheese Making

In the spring when the cows first freshened (recently having calved) and were giving lots of milk, Mother often made cheese. She filled a three gallon milk pan nearly to the brim and added a little yellow food coloring and rennet tablets dissolved in lukewarm water to cause the sweet milk to clabber. Clabbering is having the milk set similar to Jell-O or to the consistency of sour milk in the making of cottage cheese. When firm, she took a long knife and cut through the clabber in both directions as well as horizontally to make about three-quarter-inch cubes.

Once the clabber was cut into cubes, the whey began to ooze out, and it was poured off each time a quantity formed. After a large portion of the whey was separated from the curds, the curds were placed in a cheesecloth bag and hung on the clothesline to drip until most of the whey was gone. A pan was placed underneath to catch the whey, which was later mixed with bran or mill-run. This mash was fed to the chickens, and to the pigs when we raised hogs, as it was highly nourishing and an animal delicacy.

The curds were salted and placed in a gallon-sized can with both ends cut out and lined with cheese cloth. Mother cut a round, inch-thick board to fit just inside the top of the can. By putting a heavy rock on top of the thick board, she fashioned a press to press the remainder of the whey out of the curds. The press was left in place for several days until the curds became firm cheese and could be removed. She lightly oiled the outside of the cheese bricks to keep them from drying out and crusting, then wrapped them in cheese cloth and set the bricks of fresh cheese on the pantry shelf to cure. The longer they cured, the stronger their flavor.

I guarantee we never had strongly flavored or, more accurately, barely even mild cheese. Mother's cheese was so delicious we couldn't wait for it to cure. How we loved eating generous slices of it on top of her homemade bread.

Cottage cheese was a staple of our diet. A big pan of skimmed milk was set out to sour and clabber. When the clabber became stiff, it was diced with a knife for the whey to separate. Then the curds were put in a cheesecloth bag and hung on the clothesline to drain, with a bucket underneath to catch the whey. Mother then lightly salted the drained curds, added some black pepper, and stirred in some cream.

The Staff of Life

Bread making was always started the night before by first bringing up the quart jar of yeast starter, kept in the cellar for even temperature. In her gray bread crock (which I still have and use), a batter was made with the starter, sugar, flour, and water mixed to a proper consistency and left to rise overnight on the lid of the wood box in the warm corner near the cook stove. The next morning she'd take out about six tablespoons of the bread batter, in which the little yeast fungi had multiplied overnight, for the new starter. This was put back into the quart jar, a few tablespoons of sugar added, flour, and water; it was stirred, and returned to the cellar for the yeast to again proliferate until the next baking. To the bread dough she would add salt, some lard, a little sugar to feed the yeast, and more flour and water, according to the quantity she planned to bake. The bread crock was returned above the warm wood box to rise. After it rose, she added more flour and kneaded it on the table, again setting it back to rise another time or two before baking. Sometimes Mother sliced off some dough for fry bread, which we loved. Each week she also rolled some dough out oblong and thin for cinnamon rolls, which she coated with butter, then sprinkled on cinnamon and a little brown sugar, and topped with raisins before rolling it up into a long roll to cut into bun lengths to rise. At other times, she added milk and extra shortening to the beginning dough, and when it rose, kneaded in raisins and cinnamon for raisin buns. We always enjoyed a cup of tea with raisin buns. On occasion, Mother bought a bag of yeast cakes for special baking and always gave me one each day, probably because of the benefits of its vitamin B for my health. I liked them like candy.

Teas

The folks drank coffee for breakfast. The aroma being marvelous, I wanted some too, but I wasn't allowed to drink it until I was sixteen; otherwise, I was told it would make feathers grow on my back! For dinner and supper, our standard drink was green tea, which I did get to drink diluted with hot water and milk. Mother called this weak tea for children "tea kettle tea". She bought loose green tea in bulk, put a few pinches in a heavy stoneware pitcher, and poured boiling water over the tea leaves.

We did have treats of other teas once in a while. For years Mother ordered things from an herb company called Indiana Botanical Herb Company, a family business started in 1910 in which all the family helped, including printing their catalog and a helpful almanac. Ironite was a favorite healing salve she ordered and their sassafras tea that came as pieces of bark. A few pieces were boiled in a kettle of water until it was the right strength for tea. After pouring the tea off, the bark was saved and dried to boil more times until no longer flavorful.

Sage tea was another favorite, and this was drunk at bedtime. Mother boiled sage leaves until the liquid was quite strong, then added a little to hot milk. For this drink, she added a slight bit of honey. This warm sage milk, which I much enjoyed, helped to make one sleep soundly.

There was a patch of mint growing by a hydrant, where a drip kept the ground moist. Sometimes we had mint tea from the fresh leaves, but like the sage, this was usually a bedtime drink served with a touch of honey, but no milk in this tea. Grandma Brunia brought her violet leaves for tea when she visited.

Beans, Potatoes, and Onions

Many foods came in 100 pound sacks back then. The folks always had a 100 pound sack of pinto beans in the pantry. We also had dry lima beans. In the evening, Mother would take out a quart or more, spill them out on the table, and pull a small handful forward at a time hunting for little bean sized pebbles or dirt clods. Once cleaned, she'd pull these into a kettle on her lap and bring forward more beans. After all were picked over, the beans were covered with water to soak for the night, ready to cook the next day. Occasionally, a little bean-looking rock was discovered when we were eating the beans. It was hard on the teeth when biting down unexpectedly on a rock.

We had bacon and ham in our diet, so there was usually a piece of bacon or ham rind to cook with the beans. That is something we miss nowadays, for the bacon comes thinly sliced in a package with no rind. Salt, pepper, and a big onion seasoned the beans, which were a staple food we commonly had at dinner time. Another way I loved beans was in homemade chili. Mother would make up a big batch of chili, and then add some to the beans. It was great. No one's chili and no canned chili has ever tasted as good as my mother's. She learned how to make her chili from a Basque man, who used fresh ground cumin seeds and yerba buena, which gave it a unique flavor. When we went to gather a jag of wood, Mother usually left a pot of bean soup on the back of the stove to eat later. Generally, it would be lima bean soup with onions and a little bacon rind. Nothing hit the spot better than a hot bowl of soup when we came back from getting a load of wood.

The folks usually got seven hundred-pound sacks of potatoes and three of onions. These were kept in the cellar in two huge wooden boxes, one for the potatoes and the other for the onions. Fried potatoes were as basic to our diet as beans. We ate them other ways too, but my favorite way was fried. It is not as unhealthy as it sounds, as the skillets Mother used were well seasoned and required no oil to prevent sticking. Skillets were NEVER allowed to get in dishwater but were only wiped out. She cut the potatoes like French fries, browned them on one side, and then turned them over. They were generally eaten with big slices of raw onion or sometimes with dill pickles. Onions were used in stew and in many other ways, such as in wilted lettuce. At times the lettuce was too prolific to eat it all raw, so not to waste any, Mother wilted a quantity in a skillet with diced onion, pieces of fried bacon, and a bit of vinegar and salt. It was one of our favorite summer dishes.

Each fall my parents went to Penrose or Paonia to get windfall apples, as they were very cheap. The orchard people could not sell apples that fell on the ground for much money, and sometimes we were allowed to pick them up for free. This fairly long trip was an exciting excursion. We kept the apples in the cellar, and they were eaten raw as long as they would keep.

Preserving Food

The underground cellar was thirty feet from the house. The roof was slightly curved with over a foot layer of dirt on top for insulation and a four-inch square vent hole in the center. When I was small, I used to love dropping tiny pebbles down the vent. A heavy trap door opened to the eight steps leading down to the cellar, whose sandstone slab walls were lined on two sides with shelves for the canning jars and the other side with the potato and onion bins. Occasionally, I saw black and yellow spotted salamanders down there, which were scary, and rarely a snake, enough to keep us on the look-out. In all, Mother had over one thousand fruit jars in the cellar. The majority had been left in the cellar from homestead days. Many were Atlas Strong Shouldered Mason and Ball Perfect Mason, now turned green with age. The pewter colored lids and red rubber seals were reused over and over. Since we did not buy canned food, the aim was to have most of the jars full by the time winter came and to determine how many jars of each food should be canned for our yearly use. For instance, tomatoes, purchased from a valley farmer, should be nearly three hundred jars to give us daily vitamin C, as we didn't have oranges. If a tomato had a spoiled spot, it was cut away, then Mother gave it a quick sniff to be sure the rest was all right. Approximately thirty quarts of green beans were canned, which Mother snapped amazingly fast into inch lengths, to eat one jar about every two weeks. Her fingers flew as she shelled a full milk pail of peas for canning. All the jars of peaches we could get from two or three bushels of fruit were canned. Canning in those years was truly a labor of love and sweat, and I'll add endurance. It was done in the summer, when the weather is hot, and the wood-burning cook stove had to be kept stoked up to keep the canning kettles boiling. In the early years, we did not have a pressure cooker, so vegetables, such as green beans and peas, or meats had to be boiled over two hours for safety. It was a tremendous time saver when eventually my folks were able to afford to buy a pressure cooker. Tomatoes and fruits took about forty minutes per kettle-full of jars to boil and seal, since our elevation was nearly a mile high.

Mother made both dill and sweet pickles with vinegar that contained a gummy, slimy clump called "mother", which cultures homemade vinegar. The sweet pickles were made in a two gallon crock and its brine changed frequently. She chose tiny cucumbers for the sweet pickles, always so crisp and crunchy. A variety of spices, such as cloves and pepper corns, in addition to vinegar and sugar, made them irresistible.

Besides the hundred-pound sacks of potatoes, onions, and pinto beans, they also bought gallon-sized cans of honey and sorghum to store in the cellar for winter, preparing for the time when we might not be able to get to town for several months, depending on the storms. When affordable, they bought a wooden case of dried prunes. I particularly enjoyed the trips we took to haul hay, grain, and other stock feed that was stored in the barn for the cattle during harsh winter weather.

Curing Bacon and Ham

Aside from the white sow my parents acquired when I was five and a black female pig we named Tilley, whom my folks found wandering on the prairie, they had quite a few hogs a couple years later on. To use up the extra skim milk collected while dairying, they got barley and bran to make a skim milk based slop to feed the pigs. The folks always managed to butcher hogs when I was at school. I was disappointed to have missed much of what they did. With a block and tackle they hung the hog, and they had a tub of very hot water to scald the hide so the hair could be scraped off, which was necessary in order to use the rind. On the long porch table, the meat was cut up in preparation for the many ways it would be used. Bacon was cut in big slabs off the ribs, and from the sides of the jaws six to eight inch squares of meat were cut off, called jowls, and prepared like the bacon. Gallon cans of Morton's smoked sugar cure salt were used for the curing. The hams and bacon were rubbed with this salt, which drew out the moisture, and placed in a cool place to cure. The juices that ran from the meat had to be kept cleaned up. The fat was pared off, cut into small cubes, and placed in pans in a fairly hot oven to render its oil for lard. What was left after all or nearly all the oil had baked out, called the cracklings, was a delicacy to eat. All the scraps of meat were ground through the meat grinder and pepper, sage, and salt added to make sausage or head cheese. The feet were pickled.

Sauerkraut

We ate a lot of sauerkraut, especially in the colder months when there was little in the garden. Cabbage was grown, but not in sufficient amounts, so Dad brought home huge heads from a farmer in the Arkansas valley. Mother sliced them very thin, pounded the sliced cabbage thoroughly, and added salt so the juice would release. After the juice had drained, it was cured in a wooden-stay, fifty-five gallon barrel, which was put in a hole underground. When spring came with warm weather, the barrel was taken up, cleaned, and any sauerkraut remaining was canned in quart jars.

Hominy

Hominy was another food special to winter. The whole yellow corn we had for stock feed was used for this. It was boiled in huge batches in our large copper boiler. When the corn kernels were swollen and tender to eat, lye was added to dissolve the hulls off the kernels. After this process, the corn, now called hominy, was thoroughly washed to rid it of all lye. To avoid going through this process often, Mother canned all the hominy we couldn't eat soon. Hominy could be added to many other dishes or also eaten plain in a bowl with milk for supper.

Pancakes and Maple Syrup

Pancakes were generally eaten with bacon and eggs, but it was nice to finish the last one with butter and syrup. Mother made our maple syrup very simply. She just added sugar to a cup of water, brought to a rolling boil for a few minutes to thicken, then added a teaspoon of maplene or maple flavoring. Some mornings when she'd fried bacon, she took some of the drippings and stirred them into stiff sorghum as a spread for hot pancakes just lifted off the griddle. How I loved it! The old time saying "thick as molasses in January" has a truth to it, molasses being a product of sorghum.

Candy

Candy was never bought. Mother grumbled and complained when neighbors thought they were doing something nice like bringing me candy. She would hide it and then dish it back out very sparingly. During grade school, when the teacher gave a candy bar to each student who had 100 percent on their spelling tests all week, I only missed getting a candy bar one week by misspelling the word "acknowledgment" that week. I savored my candy bar bit by bit so it lasted an entire week, while others devoured their earned candy bar quickly, and the next day they would speak enviously to me because I still had most of mine. Mother used brown sugar for certain things, and I always hoped the brown sugar would have some lumps. She would pick out a lump of brown sugar and hand it to me with a smile for a treat. Sometimes she sprinkled brown sugar over buttered toast.

Once in a while Mother made taffy. Taffy was fun to make. Mother would add a cup of sorghum to the taffy recipe for flavor and cook the ingredients. She would then test it by dropping a bit off a spoon into cold water to tell when it was ready to pull. When cooled down enough to handle, we'd butter our hands so the taffy wouldn't stick to our fingers. The exciting part was when Mother and I had a light tug-of-war pulling the taffy back and forth until it got like a rope. Then we'd stop and she'd cut it into pieces.

Chocolate fudge was made at least once a year, maybe more often, usually during holiday season. Mother never had many kitchen gadgets, not even a thermometer, but she knew when the roiling boil of the fudge had reached the proper moment by the way a drop formed in cold water. If we were fortunate enough to have been given a bag of black walnuts, some were hammered on a rock to get nut meats for the fudge. After it cooled somewhat, being underfoot watching her every move, I'd see her give the fudge a vigorous whipping with a large spoon before pouring it in a pan to cool. When the fudge was finally stiff enough to cut into little squares, I was given a piece, and then a second piece. After that, like other candy, it was doled out gradually to last a long time.

I always remember how naughty and selfish I once was when someone gave me the biggest peppermint candy cane I'd ever seen. As usual, Mother took it away and gradually gave me a little bit each day, but this time she ate some of it. I remember crying for a long time, while repeating like a stuck record "You ate my candy!" At first she said she was sorry, but added that it was far too much for just me. I kept up my tirade so long that she finally got fed up, her patience ran out, and she scolded me in a stern, sharp voice, "That's enough! Stop it!" I had been far from sharing and generous. In my heart I knew I had been wrong, but it was many years later before I apologized to Mother.

The Evaporation Cooler

Without electricity and refrigeration a way was needed to keep our food cool, especially during the hot summer days. Under a tree for shade, Mother built a frame four-and-a-half feet tall by three feet wide with several shelves. To keep insects out, screen was nailed on three sides, and a screen door opened in the front. On the top board she placed a big dishpan of water and hung a burlap drape from the pan of water so that it completely covered the frame. The burlap, partially immersed in the water, siphoned water from the pan down its sides, and a breeze kept the damp burlap cool. In this way, we were able to keep our milk, butter, and cooked foods cool.

Fly Traps

There were flies galore on the ranch. Mother safely took care of them. She made traps the flies would crawl into, but didn't know how to get back out. Since the flies were in such abundance, she made her traps large to accommodate them. She formed a screen into a cylinder about the size of a two-gallon bucket; the round screen top was whip-stitched on with heavy black thread. The bottom was a six-inches high screen cone with a finger-tip sized opening at the peak, pointing upward into the cylinder, which was also tightly sewn to the round sides. Short legs were made for the trap so the flies could crawl under, and a pan of sugar water, or something similar with strong scent, set under the trap to lure the flies. They'd arrive from all sides for their feast, and when ready to leave, most crawled up the cone and through the small opening. They'd fly around in the top of the cage, but were never bright enough to go back through the hole by which they'd entered the trap to escape. Eventually they died. When the trap became nearly full of dead flies, it was emptied and started over.

Mother got rid of yellow jackets and other annoying wasps the same way. She made similar but smaller traps and used a smelly meat to lure them under. Like the flies, the wasps, also, were not sensible enough to get out through the hole they'd come in, instead frantically flying up to the screened top to escape.

Getting rid of the pesky flies inside the house was different. We had sticky ribbons that were hung from the ceiling. In those days you could buy what was called "fly paper." This was a plain sheet of black paper that was put on a shallow plate and moistened with water. Apparently, the paper was poisonous and flies eating from the flypaper dish died. To be extra tempting, a slight bit of sugar was sprinkled on the fly paper. When the dish dried due to evaporation, it was replenished with a small amount of water. And, of course, we had fly swatters. The handles of fly swatters that wore out were kept, and Mother made new screens to put on the handles, edging them with a folded, slim piece of cloth to cover the sharp edges of the cut screen.

Sewing and Altering Dresses

Of the few things Dad brought with him from Idaho when he moved to Colorado was a treadle Singer sewing machine. Whenever something went wrong, I got to fix it. I started by taking it apart, setting all the pieces out, and placing them in sequence to know how to put it all back together again after the problem was found. We always kept a small can of Three-in-One oil on hand to oil its moving parts. Mother, who was a skilled seamstress, used the sewing machine for mending and making new things. I received boxes of scarcely worn, outgrown clothes from Dad's grandchildren, all of whom were several years older than me. "Shopping" in these hand-me-down boxes was thrilling to me, having never yet been in a clothing store. Some were okay to grow into, but Mother altered many as needed to fit me. She also made dresses for herself and me from the patterned material of the turkey grain sacks. She laid out the material on the kitchen table and pinned pieces of the pattern to the cloth for cutting out. When cut, the pieces were pinned together the way they were to be sewn prior to sewing them on the machine. To make a pattern from a dress or other garment we especially liked, it was carefully taken apart, piece by piece, the pieces ironed so they were flat, then they were pinned to paper to cut around to make the desired pattern.

Mending was an on-going process, using better pieces of worn out things for the patches. Holes continually wore into the heels of our socks. To mend them, Mother inserted a small jar inside the sock and pulled the hole to be mended tightly over the jar opening to darn it. The darning thread was never knotted so there would be no bumps to bother our heels. I was quite small when she taught me how to weave the darn tightly and not leave lumps.

Old, worn out overalls made great knee patches for our Levi's, a part that invariably wore out first. Mother would cut away the frayed part, and place the patch, bigger than the hole, behind it and pin the turned-down edges of the hole to the patch prior to sewing. Her stitches were always small and evenly spaced. Also frequent were the triangular rips on the rear of our pants, which were caused by not keeping one's bottom down low enough when crawling under a barbwire fence. Cutting out the better parts from old jeans, Mother made many pairs of mittens, sizes big, medium, and small. We'd place our hands on a paper bag, and Mother would draw around our fingers to make a pattern to fit our hands.

Quilts, Heavy and Warm

In Iowa, where Grandmother Brunia lived, winter nights got quite cold, and they were also very cold at the Colorado ranch, with temperatures often falling to zero and below. Fall and spring nights were often plenty nippy too, as our house had no insulation to keep the cold air out. In Grandmother's day all the bedding used in one's home was homemade, and Mother still made all of our quilts and comforters. She once expressed her view toward quilts in a letter to a friend: "A coal fire feels good, and so do comforts at night. I sometimes wonder if there's any keener enjoyment than the quilts the mothers and grandmothers have made. It gives that extra little weight to settle around."

The quilts were always made from used woolens such as suits or wool pants. Mother made patterns for the quilt blocks from brown grocery bags, as they were fairly stiff to keep their shape when pinned to the material being cut. Big diamonds were a favorite pattern. Usually, half of the diamonds were made of several cloth scraps and embroidered with a feather stitch where the pieces joined for an artistic touch. The other half of the diamond blocks would be of a solid color, with the "prints" and solids laid out in a design.

Another favorite pattern was a hexagon approximately three and a half by six inches in area. Mother liked to use a variety of colors to make an attractive quilt top. To be different, some of our quilts had these hexagonal blocks for the top and the diamond pattern for the quilt back. When the sheep were sheared, some of the wool was always saved, cleaned, and then carded for making quilts. The carding tools, which Mother kept in her trunk, are like a rectangular wire brush, except that the wires are a quarter inch long, stiff and rigid, and spaced evenly apart. The wool must be "combed" between a set of two cards until it becomes a fairly thick, compacted "sheet". The carded wool is placed as a thick sheet between the front and back of the quilt. After the quilt is evenly filled, the next step is to get it quilted, or tacked down, every few inches so the stuffing will not be able to slip around. Mother used a soft wool yarn to tack the quilt, knotted them, and then cut off any wool yarn longer than three-quarters of an inch past the knot. There was also a binding sewn all the way around the edges of the quilts. She made slip covers about six inches wide for the top edge of the quilt so it would not get dirty from our faces or hair at night. This was attached with inch-long basting stitches, which didn't take long to do, and could easily be removed to wash. The quilts were never cleaned, but on sunny days they were hung out on the clothes line to air and fluff.

These quilts were very heavy, and when put on top of the blankets one slept in, they held in the heat. On extra cold nights it might take two quilts to keep warm. Living far from school, for a couple years I stayed with an aunt's family during the week and slept out on their back porch. In the winter, the snow sifted through the screen and landed on my bed. Some winter mornings I'd wake up with a blanket of snow on top of my quilt, but I remained warm. It definitely gave me an appreciation for the durability and warmth of these heavy, homemade woolen quilts.

I still have three of these old quilts, but one especially is in need of repair, having frayed and worn with much usage over many years. On a cot in the "roots" room, I've rescued these quilts from the upstairs closet and let them lie more comfortably where they can be seen and admired for their functional history.

Keepsake Quilts in the Trunk

Two unfinished quilt tops were kept in the trunk and taken out occasionally to work on, usually after a new dress had been made and some of the pretty material was left over. One was the beautiful Flower Garden quilt, the "flowers" set off by green pieces and one yellow piece in the center of each "flower". (In 2018, a neighbor is finishing the Flower Garden quilt for me.) The other was the Wedding Ring quilt. The "wedding rings" were set against a white background. After these many years, when taking these quilt tops from the trunk, I feel the nostalgia of remembering favorite dresses we once wore.

Making Mattresses

Mother made mattresses for the cot and the single beds. She bought a strong, heavy material, similar to a light weight canvas, called ticking, to make the mattress covers. As far as I know it only came in one color, a blue with white stripes. She used a bale of clean new straw, first shaking out the chaff and small broken straws, for the stuffing to fill the ticking, which she had already sewn, double stitched, to the shape and size needed to fit the bed. The straw was tamped down tightly in the bottom of the mattress case and gradually filled to the top, keeping all the lower straw compacted as she went. When nearing the top, she sewed the mattress case partially closed so the straw could not creep out, while continuing to stuff in more straw through the small opening left until no more could be crowded in. Finally, she double stitched the rest of the top seam closed. Mother's mattresses were quite firm and comfortable and did not sag in the middle, as she packed the straw in very tightly.

Homemade Pillows

The pillows Mother made were as soft as any goose down pillows bought in a store. When the turkeys were dressed for eating, she saved all the soft, fluffy feathers along their thighs and lower back. The feathers were kept in an airy bag, usually a mesh onion sack (onions came in hundred pound bags back then, so the bags were large), to dry and fluff. She made the initial pillow case out of ticking, same as used on the mattresses, always sewing the top nearly shut, leaving just enough room to get her hand inside the case with a fist full of down.

Feather by feather, the down was stripped off the feather's midrib and the midrib tossed aside. Once she had a fist full of down, it was stuffed into the case. Particles of down were always floating in the air, causing considerable coughing and sneezing. I helped pull down off the midribs. The down was so soft and tiny, it seemed I would never get enough to make an even slightly firm pillow. Pillow making was always a long term project, frequently left for a cold winter afternoon after the animals were all taken care of. In between working on a pillow, the small opening was always securely tied shut, as down was excellent at escaping. If enough feathers with nice down had not been saved, the pillow had to wait until the next fall, when a new batch of turkeys would be dressed for market and more bags of feathers saved.

Once the pillow had been filled with down, the small opening was tightly double sewn shut, and a second outer case was made to enclose the first case, as down feathers have a way of getting through the material. After the second case was on, also stitched closed at the top, the pillow was put in a regular pillow case, and it was now ready to sleep on. Mother's pillows were extremely comfortable and soft.

Three Braided Rugs

Before my time, Aunt Bertha's larger, one-room cabin had been sledded down the slope of her homestead land, after she married and moved away, to attach it at a right angle to Mother's cabin. We never lived in it, but used it for guests or for a living room in warmer weather. The floor was tongue and groove, four-inch boards stained brown. This floor was adorned with three braided rugs.

In years past, Grandmother Brunia and Mother had braided these rugs. They were made out of strips of old woolen pants and other wool clothes. The ends of the strips were sewn together at a slight angle and double stitched for strength. Afterwards, they were dyed to shades of purple, green, orange, and brown. Dyeing materials was common in those times. The sewn, dyed strips were rolled up in big balls of the same color to be braided. Three ends from three balls are taken at a time and interwoven to make a very long braid. The long braids were then sewn together and coiled into an ever-expanding oval shape. The same color of material would be used for several rows before changing to another harmonious color. When completed, the rugs were approximately three feet wide and six feet long.

As a child I loved these rugs, and when in Bertha's room I would often lie down on them. After a long, hot day of riding after the cattle, it was a relaxing treat to stretch out on one of the rugs to cool off. To clean them, the rugs were taken out to the clothesline and hung to be beaten with a broom until no more dust flew out of them. Today, these heirlooms of my girlhood adorn my wooden-floored "roots" room.

Not all the homemade rugs were braided. Mother often made us smaller rugs that were crocheted using large wooden crochet hooks. These rugs were made from cotton or other clothes and material that had seen better days. Like the braided rugs, she sewed one-inch strips of material together, cut on the bias at an angle, and double sewn. The long strands of each colored material were rolled into a big ball to be crocheted. She made these crocheted rugs oval in shape, starting with a straight center about two-feet long before going around and around. Her rugs were always attractive with the pleasing mixture of colors. With wear and tear over the years, none are left.

Being a busy ranch woman, Mother only did quilt and rug making after dark by the light of our kerosene lamp or on wintry or stormy afternoons. As her years crept up and her eyesight began to fade, I remember all the needles I used to thread for her when home from school. She'd have me thread a number of needles, with different weights and colors of thread, for her ahead-of-time. Mother used to say she could tell when women got over forty because they started to wear glasses for reading and sewing.

Shoe Cobbler

Our shoes needed to last a long time. Usually our soles wore out first, so Dad put on new soles, or sometimes he just recovered the worn soles. He used a metal shoe last shaped like a foot, which he slipped inside his and Mother's shoes. After shaping and gluing on the new sole, he secured it with a few short nails that were driven firmly into the shoe and clenched over the metal last so they didn't cause discomfort inside the shoe. For my small shoes, he'd find a piece of metal to put inside to hammer against to clench the shoe nails.

Sometimes our shoes needed a cobbler. Fowler, like most rural towns of that time, had a cobbler shop right on Main Street, a couple stores up from the grocery store and a couple stores down from the barber shop. I loved going into his shop. The first thing I noticed was the wonderful smell of fresh leather. He had a lively business with counters full of fixed shoes or those waiting to be fixed. The folks needed our shoes fixed the same day we came to town, so we went to the cobbler first. He had all kinds of interesting tools and a big, impressive sewing machine to sew leather.

We stopped to pick up our shoes when leaving town. I held mine close; they looked so nice mended. I felt just as proud as if they'd been a pair of new shoes. I was eager to get home, put them on, and go racing around the barnyard.

Valentines of the Past

One of my earliest valentines was from a man, a relative of the Knowlton family, probably an uncle. Mr. Knowlton was an educated man, a surveyor, and this man was educated as well. He liked Mother's lemon pies, and I for sure loved them. He sent me a valentine, and wrote on it, "I know a little girl that loves Lemon Pie." Simple, but those few words have stayed with me all these years. Everyone should take a lesson from that and tell more people a few simple but heartfelt words.

Mother was the cleverest artist at making valentines, having learned to make folded paper cut-outs when she was a school teacher. She'd fold paper into squares, and refold it several times. When cut, out would come cupids and hearts, or arrows, and lacy swirls, just the most lovely valentines in the world. Unfortunately, I didn't learn how she folded the paper to make such lovely valentine creations to be able to make them for our children and grandkids. I've tried, but never managed to do it right. Every year she helped me until I was able to make simple valentines on my own for the children at school. Exchanging valentines started in grade school and continued every year. The Krieders and the Blacks bought theirs, but the Knowlton boys made theirs too. I saved some of the fancier valentines that Ralph Knowlton made. Over the years, I've continued to make valentines rather than buy them. Only when our children had big school classes and required lots to give out did we buy packages of valentines, but they made the family valentines. Of the holidays, Valentine's Day has always been special to me, as we took the time to make valentines together. When a child, I'd add a poem to my Mother's valentine.

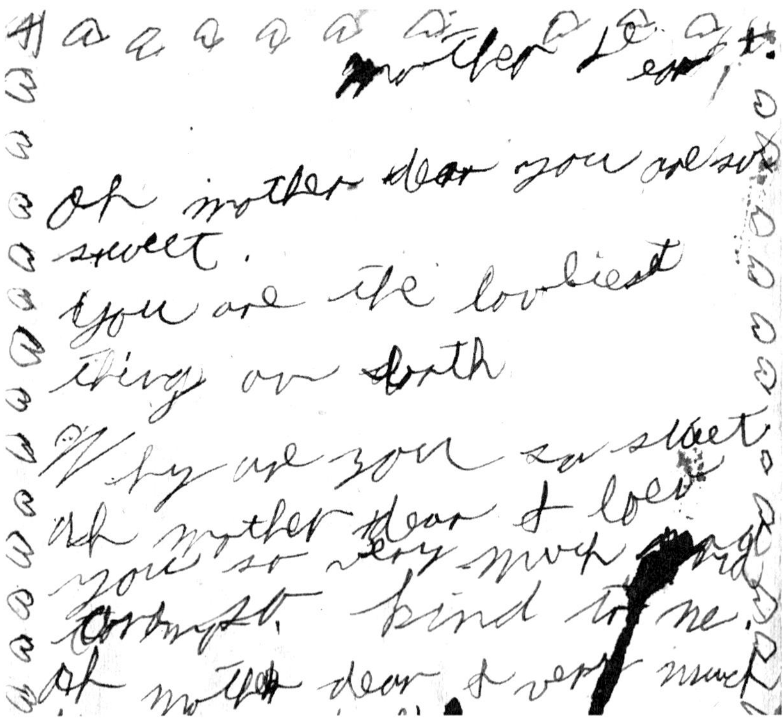
Mother Dear,
Oh mother dear you are so sweet.
You are the loveliest thing on earth
Why are you so sweet
Oh mother dear I love you so very much and [illegible] kind to me.
Oh mother dear I very much

Our Christmas Tree

Before Christmas, we always got a tree, either cedar or pinion, but the "tree" was never a tree. Mother said we should never cut down and kill a live tree; that was unthinkable for her. Our "tree" was always just a branch from a tree, and it was set in a corner in such a way that it always looked like a tree. It was exciting to get our "tree" from the cedars. Both cedar and pinion grew on some of the limestone ridges. When I was older, I used to bring our tree home on horseback. The tree was decorated with birds. Mother painted blue jays, cardinals, orioles, and other colorful birds on cardboard on both sides and then cut the birds out to hang on the tree. We used them every year. She made stars out of silver cardboard, always placing one big star on the very top of the tree. She used cotton for snow and popped popcorn to hang in looped chains on the tree. We never had a purchased ornament on our tree.

Our ranch had many fascinating places that I never tired of going to. The City Rocks, when Mother herded sheep there, was one of my early havens of exploration, where I climbed the big scattered boulders and pretended they were stores with people working there.

The canyon, an upper part of Saunders Arroyo, was a mile south of the house. It was a never-ending delight to walk up the canyon, watch the great horned owls leave the cave and fly up ahead, or to hunt for arrowheads on the rim above. The picture here shows the cave, and a drawing on the cave ceiling.

Part V

Vignettes of Growing Up on a Prairie Homestead

Riding to places off the ranch was a treat, especially down the length of the Sanford Hills, starting at the beginning just above the “Bill English”, and riding all the way, coming off at the south end. Frequently, I saw the last herd of wild horses of our area that now roamed the Sanford Hills, often around knolls at the southern end, or adjacent prairie that sloped to the east.

Haystack Butte, a landmark of the 3N Ranch and surrounding area, was a fascination and joy from the time I was quite young. I climbed its limestone shale sides numerous times to view the prairie in all directions, with its buttes and the Blue Ridge reservoir to the north, canyons and distant Pikes Peak, Greenhorn and Spanish Peaks to the west, the range of Sanford Hills to the east, and our ranch to the south with Fisher’s Peak just visible.

When native wildlife still abounded.

I am recounting the years of my girlhood, when my mother was always there as my guiding light. During those glorious years I roamed the expanse of our land and miles beyond, years when much of that part of the West was not fenced and native wildlife still abounded. I experienced such freedom; the lightness of being unburdened, unhampered by too many material things, and never knew hunger. I felt the stability of caring parents, absorbed the peacefulness of silence and space—plenty of space—and was engrossed by the ever-changing beauty of the land as the seasons emerged.

My childhood ranch memories began March 1931, when I was three, on Mother's remote homestead on the southeastern Colorado prairie. The mail boxes and nearest neighbor family were ten miles

Tom Sutton, mail carrier. Our big box is on the end.

away, the school house was then fourteen miles distant, and the small town of Fowler, accessible by a wagon road across the prairie, was thirty-five miles away. Though this life was neither pampered nor easy, the feeling of knowing one was useful and needed even as a child, a sense of belonging, of pulling one's share, however small it might be, were at the root of my happiness. Peacefulness was imbued by the vastness of the prairie; its silence and solitude; the kindness shown by parents and neighbors; the joy of new life, plant and animal each spring; and the variety of the seasons.

Every day dawned brimming with excitement and happiness, even though I had never experienced a room full of toys or playmates. As an only child, my playmates were imaginary children, and "rock families", each family being a different shape of rock, smaller rocks of the same shape being their children. I did have a few toys: a dozen blocks to build a castle, the once-a-year present of a book from Alma Sanford, a tiny, two-inch doll each year from Santa, which I called fairies. Mother made each fairy a different colored dress. Blanche Bennett gave me a Christmas box once including inch-sized farm animals and three marbles; another year she gave me *The Doings of Little Bear*, which I probably read a hundred times, I so loved each story. During the Great Depression most people did not have money to buy toys or gifts. After several years, my folks afforded a doll buggy and doll, which I loved. In time, part of the doll's face lost a few pieces. One day Mother said, "your face is breaking out." I immediately grabbed for my face, fearing it was breaking off like the doll. Mother realized my thoughts and explained it was only some rash. Another year my parents bought me a red wagon, which I pulled everywhere;

Mother made me a sled.

but Mother also found it useful. For winter she made me a sled to slide down the steeply slanted sides of the arroyo or the hill coming down to the house. When I was older I made my own beanie shooter (slingshot) from a slender, forked cedar branch, which I cut to size and notched, and a piece of rubber tubing to pull. I spent hours going around target shooting.

What follows are some vignettes of my experiences from this early part of my life. They are only somewhat in chronological order.

The Dusty Cow Trails

As a youngster around home I went mostly barefoot all summer and well into the fall when not in school. My feet were tough enough to run over rocky ground with ease. The most delightful of all barefoot pleasures was the feel of silky, soft dust between my toes. The dust in the cow paths was as fine and powdery as flour sifted twice. Our ranch soil was gray and several feet in depth, pure soil without pebbles or rocks. During the hot months, the cattle grazed far away from water when the day began to cool during the late afternoon until the following day when the sun bore down. At mid-day they headed for water single file down a cow path, which in time became a six inch, or deeper, rut with about four inches of very fine, pulverized dust in the bottom. After lying around the water hole all afternoon, the cattle would head single file back down the trail, possibly a mile or more, to where they intended to graze. One of the main cow paths began at the corrals and extended for over a mile before it petered out.

Scarcely a summer day passed that I did not take some time to cherish the feel of the warm, soft dust on my feet, especially between my toes. It was pure joy to speed down the cow paths, or at other times just to walk and roll my feet around in the velvety dust. There was one day I will never forget. Although before noon, it was already very hot when I decided to run down a cow path, speeding along kicking up little spurts of dust. I was enjoying myself so much that I paid too little attention to my feet, which were becoming painfully hot. Foolishly, I didn't turn around until the trail became faint and had scarcely any dust. Before that point, I had already sat down several times to hold my feet in the air to cool off. The mere mile back down the blistering hot, dusty path had never seemed so long. I'd run as fast as I could, then plop down, waving my feet in the air, jump up, and quickly cover as much ground as possible before the burning became unbearable again. These spurts became shorter and shorter as the heat burned deeper and deeper into my feet. Additionally, sitting on the scorching ground while attempting to relieve my feet also caused pain. I knew I was in bad trouble, with no help except getting home as fast as possible. When at last I reached the corral, there was the blessed stock tank of cool water. The folks had said the water tank was not to be played in; nevertheless, I jumped right in, to the huge relief of my feet, before running the last short stretch to the house.

Early Childhood Imagination

We lived in Mother's original two room cabin. The roof was curved with only a foot of pitch from the center to the eaves. The roof boards were tar papered, and over that was piled dirt or sod. This is why they were called "sod-roofed" homestead cabins. The roof always had a few leaks no matter how much "sod" was carried up a ladder and piled on the roof, which was mostly done in the fall to prepare for winter when the roof would get snow. We always had pans, buckets, and gallon cans set around to catch the drips. The leaks stained the ancient cedar boards that also had knots and knot holes.

When I was in the bedroom on the bed looking up at the ceiling boards, the patterns of the various stains fanned my imagination. There was a wolf; he had a knot for a large eye, a pointed stain was his ear, more stains formed his head, and he didn't have a body. There were clouds and snakes; there were buttes and a ram with a beard hanging down. The scenes changed as often as my imagination did, as I looked frequently at those old ceiling boards and the few rafters that ran cross-wise, not up and down every two feet like modern trusses.

Another thing I did when lying in bed looking up was to press a hand to one closed eye with some pressure. The outcome was that I'd see a kaleidoscope of colors: purples, lavenders, greens, magentas. It was an interesting sensation and colorful sight.

Mornings were no time for imagination. I was Johnny on the spot when it came to getting up. Some of my saddest moments were when I had slept a bit late, and the folks were already up. I'd scold Mother, "Why didn't you call me?" I'd be hurt and mad, about ready to cry, as if being left to sleep was the worst thing when I wanted to be up doing things. The folks always got up early before dawn, about 4 o'clock on the long summer days.

When out herding sheep, and all the sheep were grazing peacefully, there was time to lie stretched out on the prairie sod and watch the blue sky and the clouds drifting overhead. This is where I saw dragons, bearded goats, serpents and long necked birds, peaks and billows. Sometimes a picture would be so perfect, so real for a while, until the upper winds stretched it out, changed its shape, and it would disappear. I had so much time in those days. It's a shame I didn't have a camera to take pictures of some of the amazingly accurate shapes I saw fashioned of clouds. If the weather were just right at sunset, a lovely trim of gold edged the clouds. One could make cloaks for kings and queens from such royal materials. Not only were there gold and silver colors but glorious hues of crimson, oranges, and yellows,

Stretched out on the ground watching the sky.

fading off as the sun sank lower on the horizon into lavenders, mauves, and finally into tinted grays. Colorado sunsets gave much to spur the imagination as well as to delight the senses.

A tall, toppled chimney remained at the site where Aunt Bertha's cabin had been. It was broken into several parts. I spent many hours playing house on the different sections (my rooms) of the chimney. Mother gave me empty spice cans and other empty tins for its kitchen. I imagined I could fly, and I tried for several years to fly off the chimney using turkey wing feathers to assist me. Each time I would grasp as many turkey wing feathers as I could hold in my hands while flapping furiously and jumping into the air from this high point. Even though I always quickly landed on the ground, I was not deterred from trying again.

Haystack Butte

When past the toddler stage, whenever the sheep were grazed near to Haystack Butte (the tallest landmark for miles around due to being made of hard Niobrara limestone), I loved running up and down its front and back sides at least four times in a row, but always stopping a couple minutes to enjoy the view from this "top of the world". Looking westward toward Greenhorn Mountain, I could see an area of dark green cedars, pinyon, and pines that surrounded the rocky rims of several canyons with names like Myers, Plum, Madden, Lone Jack, and Peterson. To the southwest in the direction of Fisher's Peak, I knew that Averson, Apishapa, Bass, and South canyons were located. I had heard these names mentioned by my folks and local cowboys as places of interest to explore. When I got back down, while tending the sheep, I'd pass the time imagining what might be in the canyons.

Near the base of Haystack Butte and on adjoining low knolls there were huge, fascinating geodes. Some were broken, and their inside crystal pieces were scattered on the ground. Haystack Butte's Niobrara limestone also contains iron concretions, many of which had eroded out and were plentiful around the base. I'd always gather a pocketful of crystals and concretions before returning home.

The Arroyo

Since I didn't have brothers or sisters, and neighbor children seldom came to play, the arroyo was my best playmate, other than for the first year or two when Mother made me stick closer to the house. I spent hours playing in the arroyo with my imagination, living in make-believe much of the time. Jeffery didn't visit often to play in the early years; later on he came more often. After a rain when the arroyo had run with water, it left wet silt along the banks. Jeffery and I often played bakery and made round rolls and other kinds of breads, cakes, and cookies. That was so much fun.

I remember one time especially when Aunt Henriette's children came to visit, and Alice, Anne, and I played down on a sand bar in the arroyo east of the house. Alice was two years older than I, and Anne two years younger, so Alice was the clever one. I was so impressed that I've never forgotten the neat farm place we made. She became an artist as her profession, but she was already an artist as a little girl. I wish we could have preserved forever the farm we girls made down in the arroyo. We used little dry sticks for fence posts, grama grass heads in rows for one crop, bluestem heads in another field. There were roads in the sand, and shaped rocks for the house and barns.

There was so much to do around the arroyo that I never tired of it. I could simply go for a walk. Not far below the corrals was a long limestone slab outcropping about three feet wide and very level just above where the water would run in the arroyo during a minor flood. This was the "sidewalk", like a city sidewalk, and I walked on this sidewalk hundreds of times. The banks of the arroyo were just right to slide down, and some just right to climb up. In places there were circular hollow spots eroded back from the otherwise straight banks. These became special houses or rooms to visit and play inside. One of these special round rooms was just before the sidewalk, and I always stopped in that room. Then just past the end of the sidewalk was a gully, nearly as deep as the arroyo that headed straight north, but less than a tenth of a mile long. It ended abruptly where our road to the canyon went around it. I often played in the gully. It had steep perpendicular banks, and there was only one place at the beginning where I could climb out of it on a fairly steep trail. I liked to run down it and climb back up to run down it again.

Sometimes, when playing in the arroyo, I imagined I was a mama with a bunch of children who followed me around. They each had the prettiest names I could think of—like Rebecca, Mary, and Genevieve. I never imagined them being naughty, and I talked to them out loud as if they were really there. This fantasy likely originated from hearing Mother say she'd always wanted a dozen children, especially a noble son.

Rock Playmates

After the arroyo, rocks were my second best playmates, for I spent hours playing with my rock families around the house, mostly on the flagstone patio. The combination of rocks and imagination provided hours of entertainment over a period of several years. Nowadays, I see my great grandchildren down on their knees playing with little cars, making them drive here and there, or playing with human and animal figurines, making them do different things. So it was with my rocks. I'd gradually assembled a collection of several "families". Each family consisted of a certain shape of rock—some were round, some flat, some sort of bread-loaf shaped. In each shape there were different sizes. The biggest one was the father; one somewhat smaller was the mother; and the various sizes of smaller rocks were the children, down to the baby if there was a tiny rock in the family. Each family had a last name, which corresponded to some neighbor's name I knew. The rock family might have quite a different variety of children than what existed in the real family. My imagination had them doing many different things—they visited each other; they had homes and barns with lines drawn on the ground or twigs laid on the flagstones. They went to town; their children had games; they raised animals. As my imagination had no limit, they were never boring to play with.

Fairy Playmates

Mrs. Bennett played Santa Claus one year and gave me a few small toys and two little dolls that were two inches tall. The dolls were made of bare china, and Mother made little skirts for them out of colorful material she had in her trunk. She introduced me to the make-believe of fairies. She did a good job, for I was excellent at pretending, and thereafter I played in the imaginary world of fairy make-believe in the house. The rocks and arroyo were my outdoor playmates; the fairies were for inside. For Christmas or a surprise from town, Mother got me more of the little dolls. Over a period of time I collected a dozen of them. We made a house for them out of cut pieces of shoe boxes and lids. The fairies were not flexible with joints, so they either stood or laid down. Mother had other cloth scraps, which we put down for their beds. Many hours of my young life were spent with these fairies, dressing them and arranging things in their rooms, taking one or more out for an imaginary trip. I loved the fairies and had names for them.

Mother taught me to be truthful. But I'll never forget one of my lies about the fairies. I was around seven, and I was herding the sheep alone on a spring day in an area we called the Buttes. I pretended that the fairies had little homes out on the prairie, and that evening after I had the sheep home, I told Mother about the fairies that lived out near the Buttes in a house about the size of a shoe box. I answered all of her questions in a factual manner, making it sound as if I had actually seen them. I thought she believed me, for she never said otherwise. Looking back as a grownup, I wonder that she did not scold me for not being truthful, for she'd must have known I was giving free rein to my imagination. On the other hand, she had taught me to believe in fairies, elves and leprechauns, presenting them as real beings somewhere. Perhaps there was nothing she could say to my lack of truthfulness! One of the dilemmas of parenthood. It remained on my conscience, yet I never confessed it. I wish I had, and got it off my mind.

Tonsils

When I was nine, the doctor determined that I needed an operation to remove my tonsils and adenoids. If necessary or not, I don't know, but it was a common procedure back then. Doctors prescribed this operation for children, probably to keep them in business in lean years. I was afraid and told Mother I wouldn't go. She said I had to. I responded, "I'll run away!" The dreaded day arrived, and she got things in the old car for the trip. I did run away, or at least I tried to run away. I ran down the arroyo for about half a mile before wondering how I'd find food, where I'd hide, and if they'd find me. Finally, my conscience got the best of me, and I returned as far as the lower corral. There I fought with myself about whether or not to give up and go to the house. I went to the house, and told Mother I refused to go. Somehow she persuaded me to get into the car. Mother had won out, and we were off. At La Junta her next trial was to get her stubborn, unwilling child into the hospital and to the doctor.

An anesthesiologist came and started putting ether near my nose. I was determined to not drift off to sleep. I tried so hard to stay awake that I did wake up twice, much to their consternation, and more doses of ether followed every attempt. Finally, all I heard was the steady, low whine of the ether machine.

The next thing I knew I was in a white hospital room with Mother sitting beside me. I'd awaken and fall back asleep again. Waking was nothing like in the mornings when I was all eager to be up and out. This was a terrible, groggy feeling, topped off by feeling sick and throwing up. Mother was there, helpful and kind to me. My throat was so sore, and she gave me little sips of water.

By the afternoon, when I was able to get up, she took me to a hotel. We had a second story room overlooking the street. It was nothing fancy, just a double iron-framed bed with a mattress that sagged in the middle, a couple of chairs, and a small table. It was summer and extremely hot, and the rooms did not have air conditioning, which didn't exist back then. Mother had brought some milk for me, and something else to eat for her. That was the first night I had ever spent in town. I said to Mother I wondered how the town people could sleep. Trucks rumbled by continuously all night in the street below. It was so hot and noisy, and my throat was so sore that I only slept a little toward morning. When we got back home the next day, it was such a pleasant contrast. It wasn't too hot in our sod-covered cabin, and everything was quiet. The animals, birds, and insects went to sleep at a reasonable hour, and even the howl of coyotes in the distance happened mostly during the twilight hours. But Mother had been there, and she too suffered the heat, the noise, and a complaining child.

When Faith in Santa was Broken

I never doubted that Santa Claus existed. The day before he was to come, Mother was always so thoughtful of him. She'd make lovely chocolate cupcakes and then make a special white, boiled icing to nicely frost them. She'd put the plate of cupcakes on the clean kitchen table for when Santa would come in the night, hungry from his trip and no doubt very grateful for something to eat before he continued on his way to the other children's houses. To go with the beautifully iced cupcakes, she made hot cocoa every year. She'd make a double recipe so he could have a second cup if he wanted. Cocoa was a teaspoon of cocoa and one of sugar, mixed together with a bit of cold water. This was brought to a boil, as Mother said cocoa was a starch and not digestible unless cooked. Whole milk from our cows was added and heated, but never boiled, as milk that was boiled was constipating. The cupcakes were a special treat, since Mother usually only baked cakes and didn't make the pretty white icing.

I'd stay up as late as I could. Mother didn't suggest I go to bed until she noticed I could hardly keep my eyes open. I'd try to stay awake in bed and listen to hear Santa on the roof, but I never did. When I was a little older, I tried very hard to discover how he came and, if there was snow on the ground, search the ground up the hill and down by the arroyo, trying to pick up his tracks and the tracks of his sled and reindeer. I didn't think he stayed up in the sky when he came in the kitchen to enjoy cupcakes and cocoa and leave a present. The search was always baffling. I explained to Mother how I had found no signs in the snow. She didn't understand either, but she didn't go out looking with me.

I climbed the rock wall to look on the roof for signs; none were there either, which was really odd. I searched all around the house and outside too, but with no luck. We knew he had come, as he had eaten the cupcakes and drank the cocoa Mother left on the table for him, and he always left me something that I wanted and had, of course, never seen before. Mother hadn't seen him come either. I had total faith that he had come, but each year I wondered a little more just how he always managed to slip in and leave unbeknownst to everyone.

It was December, and I had just turned eight and was going to third grade. After school, I was at the Krieders, and Della Lee was entertaining me. She was older than me by a few months, but as far as knowing more about life and the world, she was equivalent to twice my innocent age. I suspect she knew I believed in Santa Claus, and to flaunt her superior knowledge, she happily told me that there was no Santa Claus, that he was only a made-up being. I didn't believe it, and I told her how he came and ate the cupcakes we left each year and left a present. Della Lee said it was only my parents, most likely my Mother that did those things, but there was no Santa Claus. I wondered if she could be right, but my faith went deep, so I doubted her more than believed. Nonetheless, when I was home, after a warm supper that night, I asked Mother about Santa Claus and told her what Della Lee had told me. It would soon be Christmas again in a couple weeks. I guess she hoped I could have the enjoyment of faith in Santa a little longer. Mother was tactful, neither did she lie. "Della Lee could be wrong in her information; let's see if he comes again this year. Soon he should be starting on his way from the North Pole." And sure enough, he did come, and again the white-iced chocolate cupcakes were eaten and the cocoa drunk, and a little present left.

But my blind faith was shaken; I doubted. I soon decided that Della Lee was right: there was no Santa. I no longer puzzled over how he got there when I couldn't find a trace of his tracks. Tracking was how we found lost stock. Everything left tracks, even birds when they landed. Christmas lost much of its charm and fascination when Santa ceased to come. The childhood myth was broken.

Grandma Brunia

Grandma Brunia came to visit from Iowa a couple times. She had her own ideas about what a child should learn. We'd sit for hours under the cottonwood tree, where she would demonstrate and painstakingly teach me how to crochet, embroider, and knit. She

Dad, Barbara, Mother, and Grandmother Brunia

started me on a peacock pattern bedspread, but my small fingers fumbled with skips and doubles (ways of knitting) while attempting to create the pattern. But what was worse, she thought children needed a nap every afternoon. I'd never taken a nap that I could remember; my life was too full to waste time napping. So I'd lie on the bed while she sat nearby sewing, bound and determined to keep awake, only once falling asleep to my chagrin and her success.

Grandma Brunia loved old songs and poems, and she shared them with me written out on brown grocery bag paper. One of them, which I would recite playfully to my mother, was "My Shadow" by Robert Louis Stevenson:

I have a little shadow that goes in and out with me,
And what can be the use of him is more than I can see.
He is very, very like me from the heels up to the head;
And I see him jump before me, when I jump into my bed.

The funniest thing about him is the way he likes to grow
Not at all like proper children, which is always very slow;
For he sometimes shoots up taller like an India-rubber ball,
And he sometimes gets so little that there's none of him at all.

He hasn't got a notion of how children ought to play,
And can only make a fool of me in every sort of way.
He stays so close beside me, he's a coward you can see;
I'd think shame to stick to nursie as that shadow sticks to me!

One morning, very early, before the sun was up,
I rose and found the shining dew on every buttercup;
But my lazy little shadow, like an arrant sleepy-head,
Had stayed at home behind me and was fast asleep in bed.

Grasshoppers

Grasshoppers were an integral part of my young life—a fascination. Young hopper wing-buds were mere nubbins, looking much like a pinto bean, with one on either shoulder, which expanded as they grew into adults. In flight, the fan-shaped wings of full-grown grasshoppers displayed hues of red, yellow, pink, or orange laced with a somber gray shading between the bands of color. Days on a prairie abounding with insect and animal activity were never boring to a child tending a flock of sheep.

I never tired of attempting to capture grown grasshoppers. I'd watch as one landed and closed its wings, its exterior matching the color of the soil on which it sat, then backed up and parked slightly under a few blades of grass. They always faced outward in readiness for danger. Ever so quietly I'd tiptoe from behind, slowly squat down, hand cupped for a sudden grab, only to find bare ground beneath my swooping hand. The grasshopper had sailed off a short distance for a new landing. My eyes followed her flight, then searched to find the camouflaged lady, all the while mentally honing my skills for better success the next try. Success was generally on the side of the grasshopper. When it tired of my sport, it'd fly off into the distance. Grasshoppers added an individual charm, a speck of spice, to everyday prairie life.

In the greener draws and arroyos lived an entirely different species of grasshopper called lubbers. They didn't even look like relatives of the other grasshoppers. These were big and wingless except for the pinto bean-like appendages on their shoulders. In my imagination, they became horses, since their head looked like that of a horse in spite of having no ears. In place of a mane, they wore a colored, ridged crest, or mantel, over their neck. Big, kind-looking, oval eyes protruded from either side of their head with two short feelers giving them a jaunty look. They did have one very distasteful habit of spitting what looked like tobacco juice when annoyed. The top part of their strong hind legs, which supported their weighty, plump bodies, looked just like a turkey drumstick. Their lower back legs and front legs were strong but thin with sharp protrusions along one side, making them resemble a blade of grama grass. I loved to bring the biggest, handsomest ones home to my "horse corral", which consisted of a cardboard box in the house, to play with them further.

A Budding Little Carpenter

Mother raised turkeys and made small coops for the turkey hens to lay in and brood their eggs. The coops were around 36 x 30 by maybe 36" high. I was about eight when I observed her building them. In back of the barn was an old lumber pile where Mother got her boards. I often brought home baby rabbits caught on the prairie while watching the sheep, so I decided to make a rabbit hutch like Mother's coops. There was plenty of old lumber which the folks wouldn't mind me using. I knew how to handle a saw and a hammer. It was a hot day, so inside the barn seemed like an ideal place to build it. Choosing one board to use as a measure, I laid the other boards on top and scratched the saw marks with an old nail. I did the same with the two-by-fours for the frame. A couple of buckets turned over worked well for saw horses. By late afternoon, my project was finished: a rabbit hutch with a pitched roof and all. I felt real proud of myself. Off I ran, all excited, to get Mother to see it.

The problem, which Mother immediately observed, was that the hutch was too big to fit through the barn door without taking it apart! I hadn't thought about that. As the old saying goes, "it takes the wind right out of your sails." Mother explained that the "thing" could not remain in the barn; the milk cows would need it tonight. It had to be taken apart, clenched nails and all. The only course of action was to "get busy." If Mother felt sorry for me, she didn't express it or offer any help. Perhaps she thought her child might learn a good lesson about thinking things through first.

My First Fifty Cents

One morning I was riding my horse, Ginger, over to the Sanford Ranch. I was probably ten. It was early in the year, February or March, and the snow had melted leaving the newly made Highway 10's dirt surface very muddy, especially where snow drifts had slowly melted making the mud extra soft and deep. A couple, who had braved the road in their car, were badly stuck in the mud. They had been stuck since the afternoon before. The man had shoveled and worked very hard to get out, but the car's tires had only sunk in deeper with each attempt. He asked me, as I rode up, being very excited to see anyone way out there, to please help pull them out. I wasn't too sure I could, but naturally offered to try. The man attached one end of the lariat I always carried to the car bumper, and I wrapped the other end around the saddle horn after making sure the cinch was tight so the saddle wouldn't slip. Ginger pulled her best, the man used what power the car had in low gear, while his wife pushed from behind with all her

Pulling out a stuck car.

might. Gradually the car began to budge and was pulled out to a more solid area of the road.

The man was so grateful to finally be out of the mud hole, fearing nobody would come by, and especially dreading another long, cold night. Along with his appreciation and heartfelt praise he handed me a fifty cent piece. I'd never before been paid for anything, so it seemed a fabulous amount. I was happy and very excited over my first piece of money. I remember wondering what I was going to do or get with it. I ended up keeping it in my treasure box for many years.

Neighbors never expected to be paid for each others help; it was just given from the heart. Everyone helped others whenever they saw a need or were able. Likewise, at home for all the work I did, no more thought was given to paying me than for Mother or Dad to pay themselves. Everything was just done for the sake of the family, the upkeep of the ranch, and needs of the livestock.

Jeffery Saves Me

The only playmate, companion, or brother I had in my childhood was my cousin, Jeffery, two months older, who often visited us in the summers. At six he went to school and would bring his schoolwork for me to see what was being done in school. Sometimes he went with me as I herded our flock of sheep, and I would make up exaggerated stories to be impressive. When we got into our teens, he kept a couple horses at the ranch, and at times we broke horses together. Since Jeffery's parents ran a band of sheep south of our ranch, we often rode together in all kinds of weather to search for livestock in need of care. Many times our fingers were too stiff with cold to even button a button, and our faces became sun or snow burned from constant exposure to the elements, giving us continually pealing noses.

On this occasion when Jeffery was visiting, we were riding double on Ginger with our old and only saddle. We had been checking on some cattle in the Mobly pasture, and now we were headed in a homeward direction. In this particular area, there were deep, badly eroded gullies composed of soft, decayed black shale, indicating prospects of an underground oil formation. We rode Ginger down one steep bank and were riding up the other side. We thought she could make it up the bank. Jeffery was riding behind the saddle. About half way up, with her exertion pushing out her belly, the cinch broke and the saddle tumbled off just as she made the last jump to the top of the bank. Jeffery fell off and rolled down to the bottom of the gully. But I fell off and forcefully landed headfirst in a two or three foot deep pothole on the edge of the bank. The saddle followed, falling on top of my lower back, which was sticking out of the hole. The impact of my fall stirred up so much dust in the close quarters around my head that I could hardly breathe and was suffocating. The saddle blocked off the light above me as well as any extra air.

I was dazed and had no idea where I was as I struggled unsuccessfully to get out of the hole. In the position I was in, it was difficult to get traction from my legs, and with my shoulders jammed down the hole, I couldn't make use of my arms. I thought I was a goner. It seemed like an eternity before Jeffery picked himself up from the bottom of the gully, crawled up the bank, saw the saddle lying there with a pair of legs under it, and pulled the saddle off of me. Then he had to pull on my legs with all his strength to dislodge me from the tight-fitting hole. I can tell you, I was sure glad to see daylight again and breathe real air!

Cattle and Sheep Drives

In the early years, before the land was fenced or there were any roads for trucking, cattle used to be moved in huge thousand-head herds with at least three or four cowboys riding behind and at the flanks. Usually these drives were in the fall, headed for the valley to winter, or to the stockyards to sell. Their trail was a half mile above our house and below Haystack Butte. We'd always know when the herds were going through because the dust from their passage rose in the air, accompanied by considerable bawling of the cattle and shouts from the riders. These cattle drives did not help our scant grass situation, although fortunately our land was not one of the places they stopped to graze and water the herd. For me, it was a thrilling experience watching the herd cross our land until they finally vanished from sight. In the spring another huge herd would come through, this time headed west. Large bands of sheep were also herded on the same route, headed for the valley in the fall and back to the open grasslands at the foot of the Rockies in the spring after the lambing season. The bands of sheep passing by didn't equal my excitement at the cattle drives. The Old Buffalo trail ran three miles south of our land. It was once used extensively, but each year its wagon ruts eroded deeper during cloudbursts until gullies began to form. Even so a few of the cattle drives continued to use this route. By 1940, after the completion of Highway 10, cattle and sheep were no longer trailed across land but began to be trucked. John Green Trucking was the main livestock transporter in our area.

The Old Homestead Buggy

There was an old two wheeled buggy on the ranch from earlier years, its wood gray and aged, that Jeffery and I used to take turns pulling each other around in. When we pulled it, we pretended we were the horse, but it was hard to pull until I got bigger. Sometimes I enjoyed sitting on the seat with my feet on the buckboard just pretending to be traveling around. I wished later that I'd found a way to have it brought to California, repaired it, and given our children rides with it.

Harnessing a horse for a buggy ride.

Eulogy to Bill Henderson

William Robert Henderson was born December 14, 1901, near McMinnville, Tennessee. He had a younger brother, Henry, and a younger sister, Ethel. Their mother was ill with tuberculosis and was traveling by train with her children to be with her brother, Frank Coates, in Fowler, Colorado, where she hoped the climate would help her. Sadly she died en route of a hemorrhage at the train station in Kansas City. Bill was four years old at the time, and he and his brother and three-month-old sister were taken care of by Catholic nuns until the father, who had stayed on at the family farm in Tennessee, and an aunt could come and get them.

When Bill Henderson was eighteen years old (and six feet tall with dark hair and of medium build), he came out to Colorado to be with his uncle, Frank Coates. He became acquainted with the Sanford family and lived with them for seven years while working on their farm between Fowler and Manzanola, Colorado, before becoming the cowboy for their cattle ranch. To the Sanford sisters, Alona, Alma, Carrie, and Mary, Bill was part of the family, like an older brother. They were all fond of Bill, and he likewise thought the world of them. In the evenings, with no television or videos in those years, while munching on popcorn, they'd often play games of checkers or cards like Old Maid; the girls always tried hard to rig it so Bill would come out holding the old maid card. They had great fun teasing Bill and laughing about him being the old maid.

About 1927, Bill left the Sanford's farm and became the buckaroo for their ranch southwest of Fowler, which consisted of many thousands of acres. This would become his lifetime work. The Sanford Ranch and the 3N Ranch are bordering neighbors in a vicinity known as White Rock. In the homestead years, the Sanford ranch was begun by two brothers, Jim and Bill Sanford.

The day after our arrival at Mother's ranch, I was so excited to see a cowboy, Bill Henderson, come galloping his horse down the knoll north of the house. From that day on, sighting that cowboy galloping up to the ranch house would become a familiar and ever-thrilling event of my childhood. This first visit was to welcome the folks back home and meet their little girl, but I was a shy one and managed to keep pretty well hidden behind Mother's legs.

Bill always arrived on the gallop, probably riding some young horse he was breaking. He dropped by frequently just to see people (and maybe Mother's apple pie), for being a single cowboy on a prairie miles from other neighbors can be a lonely life. It was comforting to sit down to a good home-cooked meal and talk with someone besides his horse. Everyone would sit in the kitchen visiting as Mother was preparing the meal. Soon I warmed up to Bill's persistent coaxing and spent my time on his lap. I learned to ride horses from all Bill's "horsey rides" on one of his legs or the other. We'd walk, trot, and best of all go galloping, while I laughed gleefully at the fun I was having. Sometimes he even tried to buck me off! This was the most unconditional attention I ever received, for although my folks loved me dearly, they were both reserved and too busy to expend much valuable time for play.

The folks and Bill never ran out of things to talk over: the cattle and calving problems, drought or cloudbursts and snowstorms, the range land or the scant grama grass, neighbors, coyotes or grasshoppers, prairie wildlife, trap lines, arroyo floods and water gaps, tumbleweeds or huge rattlesnakes. One time when Bill was fixing fence, there was one of the big fellers coiled up for shade under a tumbleweed that had blown against the post he was stapling the bottom wire to. When the rattler loudly made its presence known, Bill jumped so fast he almost sat down backwards. Politics and world affairs were also important topics of conversation. I remember they didn't trust the Chinese and thought that China might war against us in years to come. And yet amid all this serious talk there was plenty of laughter as well. Bill was good at jesting, and he'd laugh over his own jokes hardest of anyone. He had a hearty, unique laugh that I can still hear clearly yet cannot describe.

Bill Henderson had a little black coupe he drove when he went into town, not often. He was paid $30 a month for punching cattle, which was standard pay in those years. But that was more money than my folks had. Bill always remembered his neighbor child, and after he'd been to town he brought a few pieces of hard candy in his pocket for me. Mother didn't want me spoiled, so she'd object, but Bill always insisted, "Oh, a piece or two won't hurt her." One year around Christmas time Bill came galloping in with a bundle under one arm. He handed me a stuffed Mickey Mouse, over a foot tall, with velvety red pants and a white shirt. I loved it so much I took it to bed with me every night. With Mickey cuddled in my arms, Mother could see his big bright eyes looking up at night. This was the only stuffed toy I ever had. Mickey is now retired to the attic as a keepsake in memory of Bill, who believed that every child deserved a toy.

My parents had a little flock of sheep, and I became a sheepherder when they considered me dependable to herd sheep alone. Bill was a cowboy, definitely on the side of the cattleman, and he would tease me unmercifully time and again about being a sheepherder. He took great delight in reminding me of all the feuds and wars between the sheep men and the cattlemen, and how cowboys would rope sheepherders and drag them, which sounded very dangerous for the sheepherder. He'd keep such a stern face, so I was pretty sure he might really mean it! When I grazed our sheep close to the Sanford fence, a three-strand barbed wire fence with a high bottom wire for cattle, our sheep could very easily crawl under. And they had to sneak under because in the eyesight of sheep, the Sanford grass was much greener and better than ours. When I had been careless, the big telltale problem was they left bits of wool off their backs on the bottom barbed wire. To my dismay and trepidation, Bill unfortunately always quickly noticed! Next day he'd be over with a grim look on his face and another frightening episode of what he might do to the sheepherder.

Bill was a knowledgeable trapper and successful during the cold winter months when the bob and lynx cats, coyotes, and badgers had thick fur. He trapped all over, but he caught most of the big cats up west around Myers Canyon. He'd stop by our house on his way home with the animals slung over the front and back of his saddle. I'd stand by his horse looking up with awe at how big they were and reach a hand up to stroke their soft fur. Bill always carried a lariat on his saddle, but these times he also had his rifle in its leather scabbard along. Bill made stretchers to fit each size and kind of pelt. His place was a child's paradise with all the pelts he stood up around his porch and in the house to cure and dry before selling them. My parents trapped coyotes and made stretchers for their pelts too, but I delighted in seeing all Bill's furs lined up, for he had so many more.

Then I became eight. I tagged along about as close as a noon-time shadow when Bill and Dad walked around outside, talking. I had graduated from Bill's lap to his side. We were becoming cattle people, not just merely sheep people. Somehow having cattle as well as sheep made me feel more important in Bill's eyes. At least he did not tease me about our cattle.

The folks bought a twelve-year-old mare, Ginger, when I turned eight. She'd been a true cow horse in her day, and hadn't forgotten a thing! She taught me how to punch cattle, get them rounded up, or how to cut one or two head out of a herd. She was quick, smart enough to outwit any cow, and certainly didn't need any instruction from me. My other instructor was Bill Henderson. Now I could ride horseback with Bill, and my horizons widened from the home ranch to the huge Sanford range, exciting new places for any youngster.

In those years there were quite sizable bands of wild horses ranging in the Sanford hills and on the prairie to the east and south. Ginger was a barren mare, but still she escaped several times to get with the wild horses. Whether their stallions had lured her or she just wanted horse company, I don't know; neither did I discover how she had escaped our fenced east end horse pasture. I would walk many miles to locate her, and then was unable to catch her. Bill would help me get her back. He'd put me on one of his safest horses, although still plenty spunky, and we'd ride out down the Sanford hills to find the herd she'd be with and chase it north down the hills. These times were so exhilarating racing after these wild horses, manes and tails flying in the air, clouds of dust sailing with the wind. There was always a lead mare in front, and generally the stallion guarded the rear, ready to fight for his harem. That part could be scary, and Bill warned me not to ride in too close. As the horses neared the end of the range of hills, we'd veer the fleeing band west off the buttes, where they would become gradually funneled into the Sanford's large, round, high cedar picket corrals before they realized they were being trapped. There was such a commotion with all the snorting, rearing, and loud neighing, as this truly wild, terrified band raced around the corral looking for escape.

Barbara with pet antelope, Baby Doe.

One day Bill rode over with a real surprise in his arms. He'd found a baby antelope whose mother had died, and he brought it to me for a present. It was almost all legs on top of stilts! Baby Doe was so sweet. My mother, being a nurse, figured out a milk formula that agreed with her after a few tries. As a baby she would only sleep on our bed, no floor or corner for Baby Doe. True to her heritage, she loved to run. I spent hours racing Baby Doe around the corrals. She'd run with me to get the milk cows in the evening, but never seemed to accept that we had to walk them home slowly. As a yearling, she ran off one afternoon to go back to the wild, never returning.

When the Sanford cattle needed to be moved from one pasture to another, Bill would come over and ask for my "help". Whether the day was a hot, sweaty, dusty summer one—a foot freezing, nose-running zero winter one—or just a plain pleasant spring or fall day, every ride with Bill Henderson was a highlight of my childhood. If we were moving cattle from their north pasture to the east pasture, we'd ride to the far north end, then each take a side of the nearly two mile wide pasture, Bill to the east with fairly flat land, and me to the west with some knolls and draws requiring a careful search to not miss any small grazing bunches. We'd bring the cattle we found to the center of the pasture and gradually keep the herd moving south as we kept adding more bunches from either side until we had all the cattle in one herd of abou 400 head, then turned them east into the new pasture. I may have helped some, but then I only thought I was having a marvelous day. Before I rode home he fed me a hearty meal, such as beef stew he had cooked previously, with a cup of black coffee. Mother had said I wasn't to drink coffee, but I never told Bill.

Some days with no serious cow work to do other than checking to see how the cattle were doing in the different pasture areas, we would stop at someone's abandoned homestead cabin and look at the remains of an old cook stove, iron bed frame, or a pack rat's big pile in one corner with dish fragments, letters they had left behind and old papers mixed with cactus bits and cow dung. A cistern where they had once stored their water would be near the door. One homestead I loved to visit was a dugout near the end of the Sanford Hills with parts of an old broken wagon scattered about. Bill told me interesting stories about these adventurous settlers whose homes these had been a decade or two before.

Coyotes and jackrabbits were plentiful as were the antelope. One day when we were near some antelope, Bill got off his horse and asked me to also dismount and lie very quietly flat on the ground. He held an arm straight up with his hat on it and jiggled his hat up and down. The antelope got so curious about this "thing" moving in the air that they came over very near our horses before they saw us on the ground and bolted in surprise.

Bill Henderson always managed to put excitement into our rides, or fill in with a challenge to my mind. For his few years in school he was well educated, especially in history and math. Bill could add a column of three digit figures like an adding machine and then wondered why I couldn't do likewise with all my schooling. I needed a paper and pencil. No way could I add in my head and come out correct like he could. He truly amazed me. I've used those horseback math lessons of Bill's to this day to at least recognize if totals are approximately correct and logical.

The years of being between twelve and sixteen came. Now I was a real cowgirl. I had Laddie, the horse of my life. He was lean, strong, had supreme endurance, and was oh so fast. I trained him for cow work just as Ginger had trained me. Now with Laddie I could truly

Laddie, the horse of my life.

help Bill and give him a full day's real work. Bill complimented me many times, saying that I was better than any man that ever helped him. I did have good cow sense, but just the same it was very typical of Bill, as he never was too complimentary of other men, to the point of even being critical of most. On the other hand, there were very few women that Bill did not have good words for and what great workers they were. But his greatest love and devotion was to the Sanford sisters: Mary, Carrie, and Alma (Bill always said Al-me with his Southern accent), who received his highest compliments and praise.

Bill Henderson broke many horses in his day, chosen usually as two-year-old's from the wild bands. After he'd got them to lead, he would cinch an old saddle on them, and tether the bronco to a heavy log where there was grass not far from the corrals. Other than leading them down to the arroyo for a drink, he'd leave them tied for a couple weeks before beginning to ride one. He kept a string of some eight or ten saddle horses he'd broke so there would always be fresh horses to use, or for the Sanfords and other cowhands, who came at branding, weaning, and other round-up times. One of the thrills of my life was being invited to these round-ups. Alma (and Mary in earlier years before she married, and Carrie, too, sometimes) brought quantities of great eats and would get the food ready before they joined the riders. Aging Uncle Bill Sanford (who lived to 101) rode, a bit hunched over, to oversee the whole operation. For many years, he applied the branding irons, the familiar + HH.

Bill had a favorite horse, Smoky, he often rode. Now he said that Smoky could outrun Laddie. I knew better and challenged him to a race to prove it. Every time I wanted to have the race, Bill always had an excuse to put if off. I figured he knew Laddie would win for sure. But he said he didn't want me to feel bad when "my sorry old plug" lost! We teased back and forth as to which of us had the best horse, Bill bragging up his Smoky and running mine down. I knew it was Laddie, of course. At least this way it never was proved who did have the better horse. But Bill never missed a chance to give a dig about my "sorry" horse, enjoying the loud protest he always got out of me.

During these years, Bill started to tease me about boyfriends. Bill was in his glory if he could tease someone into a quarrel. I never had a special boy friend, but if he found out which boys I liked, or if the young boy came over to see me from the Redtop, he'd think of all kinds of reasons why they were worthless, no good, and lazy. I don't think he was teasing about boys, but it was just the way he felt about competitive males of any age. Bill hated false things, make-up included. I started to wear lipstick during my teens, and he really hated that. He said it made me look painted and horrible, that I'd be old before I was twenty, that he ought to rub it off my face. He teased me so bad that if I saw him riding in, I'd run to wash any make-up off my face in a hurry.

High school years came and went; now was college. I never failed to ride over to the Bill English, as everyone called the Sanford's cow camp and house on the hill, and see Bill each time I came home. He would cook a great-tasting meal, and we talked about what had happened around the ranches while I was away, and about college life and learning. If time allowed, he invited me for a ride down the Sanford hills, always one of my favorite places. Things were changing. Bill wasn't quite as slim and agile. He had a pickup and was doing a lot of his cow work driving rather than on a horse. In later visits, we took pickup rides over the Sanford Ranch to see how the windmills were doing, if the ponds and reservoirs had water, how the cattle looked, if there was still salt out near the watering places.

My married years were coming. It was time to make a change. But the thought of giving up ranch life, vast space, blue skies, the horses and livestock I loved, prairie neighbors, and especially this different kind of freedom that I would never again know, frightened me. I rode off on Laddie early on my last morning of single life (June 18, 1950) to try and conquer the turmoil of my inner battle, even considering that I just might ride away and hide in some canyon. I was headed east and Bill happened to be out on horseback and saw me. He galloped over, and tearfully I expressed my fears of leaving the kind of life I'd always loved for one unknown and vastly different. Bill's eyes spanned far that June morning, over the prairie range land of which he seemed such an integral and natural part, then turned his gaze east toward the Sanford hills he loved, and understood so well my parting agony. A heavy stone filled Bill's chest that morning, knowing the young cowgirl companion that had grown up in his shadow these past eighteen years was being pulled from his side, leaving a hollow spot that would be slow to fill. He lent an attentive ear as he listened, but he didn't try to persuade me one way or the other. He did offer to bring me food should fear take the upper hand.

I thanked him then headed off to ride alone down the Sanford hills. Bill wheeled his buckskin north toward the Blue Ridge. I rode for several hours in other favorite haunts as well when big, black, ugly afternoon clouds came quickly over the Rockies from the west. I knew a bad storm would hit soon, and I turned Laddie's head for home. He again raced like the wind, covered miles in minutes, just in the nick of time for the wedding and to an extremely nervous groom! The minister was there, Bill was there, family and a few friends were all waiting. I made a dash to the bedroom, kicked off boots and jeans, and a few minutes later came out in my wedding dress.

After I was married, Vinson, the children, and I would go to see Bill when we were home for a visit. Now he was using only the pickup to take care of the Sanford Ranch and was beginning to complain more and more of his hip hurting and going bad. He had his hip operated on, but he died in the hospital in July of 1972. Alma Sanford phoned immediately when Bill died and asked if I'd write his eulogy. I sorrowfully accepted, as he had given me some of my richest memories and best times of my young years. In Bill's isolated life I had been a sunny presence sharing many rides and a silent listener to the recollections, dreams, hopes, and plans of this honest, good man. We will all miss you, Bill Henderson.

A Cowboy's Prayer

Oh Lord, I've never lived where churches grow.
I love creation better as it stood
That day You finished it so long ago
And looked upon Your work and called it good.

Just let me live my life as I've begun
And give me work that's open to the sky;
Make me a pardner of the wind and sun,
And I won't ask a life that's soft or high.

Make me as big and open as the plains,
As honest as the horse between my knees,
Clean as the wind that blows behind the rains,
Free as the hawk that circles down the breeze!

Just keep an eye on all that's done and said
And right me, sometimes, when I turn aside,
And guide me on the long, dim, trail ahead
That stretches upward toward the Great Divide.

—Badger Clark

Alma Sanford

I owe much to Alma Sanford, and hopefully I expressed some of my appreciation to her before she passed on. Alma never married. She had dark hair, nearly black, and lived in Manzanola at their headquarters. She often came out with her uncle, Bill Sanford, to check on their cattle, ranch land, and their buckaroo, Bill Henderson. Her uncle, Bill Sanford, had never married either, and he lived to be one hundred. He could still ride, albeit hunched over, nearly to the end of his life. He had his ways, shy around the ladies, easily embarrassed, and he could never say a word like "bull", which he considered improper.

At branding time and during roundups, when the Sanfords took hundreds of their cattle from the ranch here across country to their Manzanola farmland pastures to winter them, Alma did her full share of helping out. The men would be on the trail for a few days (with a long, broad string of cattle and lots of dust), so it fell to Alma to keep everything organized and see to it that hearty meals were cooked and brought to the men trailing the cattle, consisting mainly of Bill Henderson, her uncle Bill, Harry Noble, and a young man or two. She'd bring a lot of food already prepared from Manzanola as well as cook at the cow camp. She was the one who kept the books. I don't know the ownership of the Sanford Ranch besides Alma, her sisters, and Bill Sanford, but only Alma took on the day-to-day work for her lifetime.

When they came out to their ranch, Alma often dropped in at our ranch for a short visit. Alma was always so neat and tidy, usually in a trim, clean cotton dress, never a flouncy one. She was an exemplar in my young life, and later on as well. When Alma came to visit Mother, she always talked to me as well, making me feel important. When she dropped by, she'd bring some food to share. Every year at Christmas time, she sent me one book, beginning with *The Alphabet Book* when I was a small child, and on up to *Treasure Island, Robin Hood*, bird books, and *Alice in Wonderland*. I reread the books that Alma gave me dozens of times, since they were practically the only children's books I had growing up. When I was older, I read *The Girl of the Limberlost*, which Mother had, many times over. Mother only afforded buying me one book when I was young, a dictionary for young readers. Alma will always be treasured for the books she gave me, my start at loving to read. I did have two school readers after I was six, one for first grade and another for second grade. When I started school in third grade, there were books to read at school. The school had a library—a five-shelf bookcase—along the south wall. Mrs. McKinney, our teacher, managed to add several books each year. I read every book there, most several times, according to my age and grade.

Alma wrote to me when I was away at college to ease my homesickness, and she continued to write to me after I became married and moved to California, keeping me in touch with the ranch, neighbors, and Fowler people that I had known. I'm thankful to have visited her in Manzanola not too many years before she passed on. Alma was a true and loyal friend to me all her life. She was the one who wrote to tell me about my mother (who didn't want me to know) that she was in the hospital with a broken hip, caused by a bale of hay falling from the top of a truck load and knocking her down. Typical of my nurse mother, she had instructed those taking her to the hospital to first make a splint and wrap her leg with strips of sheet to keep it stabilized on the long trip to Pueblo. Although I was in California, I was having nightmares of Mother being hurt. I knew something was wrong, but what? I was so thankful to Alma for not listening to Mother and letting me know what had happened. Even though I wasn't able to come home then, at least I could write to her often, as she was in the hospital a few months. I sent some sheepskins for her back to ease her bed soreness misery.

Schools

In the early homestead years, there were eight schoolhouses within a 20 mile radius of the Three Nurses ranch, a testimony to how many families attempted to homestead in the area, but as the hardships of homesteading became too much, many families sold their land and left. By the time I attended school in 1937, only the White Rock school remained. It was originally located about two miles south of Highway 10 and one mile east of the White Rock Road, but in 1939 it was moved four miles to its present location.

The names of the other schools that had been abandoned by the time I went to school included the Eureka school; the Mustang, northwest of the Mustang Headquarters; the Clanda, north of the Hart Ranch headquarters; the Stony Point, just west of the mouth of Apishapa Canyon; the Prairie View, several miles northeast of the Red Top headquarters; the North Omar, a couple miles south of the Mustang arroyo, just west of the Fowler Road; the South Omar, somewhere north of Highway 10 on the east side of the Fowler Road. East of the Mustang headquarters was a pasture called the schoolhouse pasture. This may have been the location of the Eureka schoolhouse, which was moved to attach to the White Rock schoolhouse.

White Rock: a One-room Schoolhouse

When a child turned eight it was mandatory in Colorado to attend school, home schooling after the age of seven being against the law. Mother had kept me home as long as possible. The school was fourteen miles away: bringing me there and getting me meant fifty-six miles of driving a day over poor dirt roads and often in stormy weather. This was time and expense my parents could ill afford. Mother had taught me arithmetic and spelling as we herded sheep, and reading and writing at the kitchen table. Now eight years old, I entered the third grade in September 1937.

The school attendance, divided among the eight grades, averaged eight students. For two years, a family with four children moved nearby, and then we were twelve students. My class was the largest with three pupils: Bob Black, Della Lee Krieder, and myself. Mrs. Reva McKinney had been the school marm for years before my time, and she continued to be one for years afterward as well. Although she had no children of her own, she was devoted to her pupils, giving each an exceptionally well-rounded start for high school and beyond.

Reva McKinney

Mrs. McKinney maintained strict classroom discipline and never showed any favoritism. Besides being a marvelous teacher, she must have been an expert driver, since she drove back and forth from her home in Fowler, twenty-five miles away, over roads that were mere wagon trails in her beginning years. The ruts were muddy and snow-covered in winter and spring, but never once while I was in school was she not there ahead of the kids, with the pot-bellied stove warming the room when we arrived and the flag waving atop the flagpole.

After a couple years our schoolhouse, a white building about twenty by forty feet, was moved closer to the Knowlton and Krieder homes, and four miles closer for us. That same summer, 1939, a second, old, white schoolhouse, the Eureka, was moved on skids from up north. It was attached to our school to make a second room for the teacher to stay in during bad weather. Moving the other schoolhouse in one piece across the prairie a distance of about eight miles was some feat. It was too heavy for horses to pull alone. A post was set some yards ahead, and with the help of horses and a block and tackle, it was pulled up to the post. The post was again set farther ahead, and the same operation was repeated time and again. Merle Black was one of the men who helped with this undertaking.

The school had two outhouses, one for the boys and another for the girls, with a beaten path leading to each. My first day at school,

Mrs. McKinney told me the school rules. One rule was to hold up your hand for the teacher to give permission to leave one's desk. I needed to go to the toilet, so I held up my hand. Soon I had to go worse and worse while holding up my hand at arm's reach, and still Mrs. McKinney didn't acknowledge me. Finally, I could hold on no longer and I wet my pants. It ran down the back of the seat and onto the floor. I was a very shy child, and this was extremely embarrassing for me, especially on my first day of school. When Dad came to pick me up after school, the first thing he said was, "It smells like you wet your pants." I had never wet my pants since I was a baby. When I got home and was talking to Mother, I used the word "okay". She thought that even on the first day of school my use of English had gone downhill, as we said "all right" and never used "okay".

The way Mother insisted that I dress became an embarrassment once I was at school, as the other two girls dressed much more "modern". Mother cared nothing for dress style, compared to dressing for health. I had to wear long, brown cotton stockings held up by her homemade supporters. It was always a big relief when spring weather became warm enough that Mother said long underwear and stockings were no longer needed. To support young ankles (so Mother thought), I wore shoes that fit above my ankles—never pretty low shoes. My dresses were made over hand-me-downs and printed feed sacks, which I had thought were beautiful until others made fun of me.

White Rock School, Reva McKinney, and pupils.
(l. to r.) Ralph Knowlten, Dorothea Mumm, Barbara Black, Mrs. McKinney, Della Lee Krieder, Vivian Welch, Jeffery Black. Front row: Nelda Fay Welch and Nadine Knowlton.

I liked the other kids, but I didn't quite know how to make friends having been raised a single child, whereas all the others came from families with three or more children. Living miles from their families, I had seldom seen or played with them, except for the three Knowlton boys during the few times their family came to visit. Their mother, Tressie, and mine loved to visit and giggle like two girls whenever they got together. Play is an ice-breaker, and before long I began to have fun at recess and noon. There were two swings, a gym bar, and a teeter totter. We played marble games (digging little pits to try to knock the marbles into them) and jack stones, both played on the ground. There

Playground swings.

were hopping games, such as hopscotch, with squares drawn on the dirt to jump on or skip over, sack race, three-legged race, kick the can, tug of war, many kinds of tag games, and Red Rover for running. We played Andy-I-Over (we said Andy, Andy Over) frequently by throwing the ball over the schoolhouse roof, hoping the team on the other side wouldn't catch the ball. If the other team caught the ball without it hitting the ground, they rushed around the building to tag our team, who all dodged their best to run to safety on the other side without getting tagged. If tagged, you were out. If the ball was not caught, it was thrown back over to our side. (Few schoolhouses are small enough nowadays to throw a ball over the roof.) Baseball was my favorite game. As there were too few of us for teams, we played "work up." Winter snows were great for rolling huge snow balls to build forts and then have two opposing teams attack each other.

Hide and seek was always popular as well. The school had a barn for the horses. One windy day, as we were playing hide and seek, I ran to hide in the barn. Fancy, Della Lee's pinto horse, was startled when I dashed behind her. With the wind blowing hard, she hadn't heard me coming. She flung out both hind legs, and one hoof struck me on the side of the head. We didn't consider a doctor or stitches back then, so the gash was pulled together with "butterflies" of adhesive tape.

After recesses and at noon, Mrs. McKinney rang her long-handled, loud-sounding copper bell, and we all quit playing immediately and ran for the schoolhouse like a covey of little quail. Everyone had been taught at home to mind instantly. Our school did not have electricity. Drinking water was hauled in a tank and poured into a cistern. A bucket of water was bailed up as needed and a dipper hung over the side. We all drank from the same dipper. There was an enamel wash pan for our hands. The horses had to wait and drink at home. The water bucket was in a little entryway of the main schoolhouse, where there was also a pile of coal for the stove. Steps led down to a basement where the year's coal, a bit of wood, and kindling were stored. This entryway was also where we lined up our lunch pails. The top part of our lunch pails held a thermos of hot soup or a drink. A row of hooks was attached on the wall for our coats, and our rubber boots were lined up below on the floor.

Teasing and playing pranks happened quite frequently. One time Ralph told me he'd give me twenty-five cents if I'd come to his place before Dad came to pick me up. I wanted twenty-five cents. As he walked me around outside his house, he asked me to smell different things. Soon Dad came and I asked for my twenty-five cents. He said he'd already given them to me, twenty five things to smell—scents! I was really disappointed, besides falling for his trick. Once Della Lee found an immense bull snake, a good three feet long and large around, just the thing for her to dare me to take it. I accepted the dare, took it, and to prove to her that I was not afraid, hung the snake around my neck, like I had seen the boys do. I failed to hold its neck close to the head, and it being very strong, managed to strike me on the lips. Even though non-poisonous, evidently he'd eaten something bad, so my lower lip stuck way out, badly swollen. At home that evening, Mother was most disgusted with me.

I was in seventh grade when a new family moved into the area. Two girls were in grades below me, and their tall gangling brother of fourteen entered eighth. It wasn't like me to dislike people, but this boy had teenage pimples and was repulsive in my estimation. Worst of all, he was always trying to get a hold of me, and nothing could be more terrible. I would maneuver away in the schoolhouse when he tried to get close, but the playground proved a daily battle of wits. I had to stay ever on guard to keep from being caught and kissed. One day at recess, he was chasing me. Our schoolyard was a few acres in size to allow for the horses we rode to graze. The back part had quite thick brush, some three or more feet tall. He was long-legged and fast. Mine were short, but also fast, yet he was rapidly gaining on me. Each way I desperately tried to outmaneuver him to get back toward the schoolhouse, he blocked. Terror shot through my chest. Immediately, a strange feeling came over me, one I'd never experienced before or afterward. My sheer panic left, and sort of a fluid, warm calmness enveloped me while running faster than I had ever run before. I sailed over the ground and took flying leaps over the tallest bushes, quickly outdistancing this horrid boy. In seconds, I was back in the safety of the schoolroom. He never did manage to catch me.

All the other school children rode horseback to school except for me at first, because I lived so far away. The three Black boys, Buddy, Jim, and Bob, who lived only about a mile and a half from school, all piled on a tall, slim bay gelding. The Knowlton boys also rode about a mile and a half to school by cutting diagonally through their sagebrush pasture. They all crowded on old Bulldog, a sway-backed, short-necked, dark brown horse, Allen in front, Harry in the middle, and Ralph hanging on the rear. Della Lee rode Fancy, a small black and white pinto, and Frances Krieder had a tall brown horse. They rode the farthest, a mile up to where the Blacks lived and then on down to the school. By fifth grade the Blacks had left, and Jeff and Aunt Sara and my cousins, Jeffery and Alan, moved into their place. I stayed with their family for two school years. At first Jeffery and I rode horseback to school double on his Ranger. Then we brought Ginger from home. Mother walked along beside me as I rode Ginger to Aunt Sara's place, going directly across the large Sanford and Krieder ranches and coming out by the sulfur well from which Aunt Sara's family hauled their drinking water.

Each morning we got our horses, took a lunch, and rode east to the school. There was a shed for the horses, and they generally stayed around it for the day, not grazing much, as the schoolyard grass was scant. Saddles and bridles were left in the shed. One day three of us were racing our horses down the road. I was riding bareback on Ginger, who was just flying in the lead. She stopped on all fours at the gate where the Knowlton boys generally came through, and I went sailing over her head, landing on the ground far out ahead of my mare. Luckily, I was not injured other than a few scrapes and bruises.

When the schoolhouse was moved, Jeffery and I rode the other way, north past the Krieders. The other school children now walked. I remember vividly the early morning winter rides when the temperature was zero and below, the snow making crunchy sounds under Ranger's hoofs. Our feet would get so cold riding that we'd get off to walk, stomping our feet for circulation and warding off frostbite. At that time, I didn't have my own horse and rode double behind the saddle on Jeffery's horse.

In the seventh and eighth grades, I stayed at home and sometimes rode Laddie the ten miles to school, having to leave very early to get there on time. With Laddie, a fast trotter, trotting steadily for ten miles, I thought I'd be to school on time, but even so I did arrive late a few times. Mrs. McKinney was strict and sternly let me know when I was late. None of the other children were tardy or late—even with horses to catch and saddle, bad weather, having to do chores at home, like milking cows, then riding a couple miles (not ten like me) to the school. Nowadays kids have it so easy, and they are tardy more often than were our one-room school ranch children.

When the weather wasn't appropriate for outside games at recess or noon, Mrs. McKinney had us singing around the piano. She played for dances and could play very well. Usually, one period later in the afternoon was devoted to singing. One of our favorite songs, among others, was "Frog Went a Courtin' and He Did Ride." We also played blackboard games like tic-tac-toe, hopscotch in the aisles, jack stones on the floor, and made wooden or paper crafts at our desks. One of my favorite inside games was playing musical chairs at our desk seats. Other indoor games were treasure hunt, London Bridge, Steal the Bacon, I Spy, Pick up Sticks, Simon Says, and string games on our hands. Sometimes Mrs. McKinney read an exciting story to us. She also taught handicrafts. The boys had coping saws and made things out of plywood to be painted. I still have two things Ralph made: a green letter holder with a white daisy painted on the front and a square hot pad with a bucking horse. The girls were taught to knit, and we made a few nice sweaters to be sent to the Red Cross. In art, we made pretty posters she put up above the blackboards, mostly of flowers, but the Knowlton boys usually had horses or cowboy drawings on theirs. Some afternoons Mrs. McKinney taught us dances: the waltz, polkas, but best of all square dancing. Once she saw that we had the sequences memorized, she'd play the square dance tunes on the piano and give the calls. These lessons came in most handy when dances were held at the schoolhouse. At the end of the day, just before going home, she gave us individual piano lessons, two lessons a week per student in the upper grades. I am very grateful to Mrs. McKinney for this early start at music. At that time Mother found an excellent piano she got cheaply from an elderly lady who could no longer play, but was happy to have her beloved piano go where it would be appreciated.

Our desks were arranged in rows. There was a long bench in front of the classroom by the wood stove, near the teacher's desk. Mrs. McKinney would call up the classes to recite lessons on the bench, such as: reading, arithmetic, spelling, grammar, history, health, and geography. Penmanship was done at our desks, which had a hole for an inkwell and a quill pen. Sometimes we wrote cursive on the blackboard. The lower grades were always hearing the upper grades recite. This gave us a good head start on learning. One of my favorite events was the spelling bee with all eight grades. We'd all stand in a row, and Mrs. McKinney would give the first person a word and continue down the line. When someone misspelled a word, they were out and the next person had a try at the misspelled word. The winner was the last person left standing.

There were special events. Twice Mrs. McKinney took us to Rocky Ford to see a movie. Snow White and the Seven Dwarfs was the first movie I ever saw, and later, The Wizard of Oz was my second movie. These experiences were amazing, like magic, considering we didn't even have a radio at home. We would all huddle in the back of a pickup under a tarp on these trips. Every spring near the end of school we had a picnic at one of the canyons, often Meyers Canyon.

A spring school picnic at Indian Ceremonial grounds near Apishapa River (l. to rt.) Jeffery Black, Ralph Knowlton, Della Lee Krieder, Dorothea Mumm, Lois Hamilton, Alan Black, Lucille Huntley, Barbara Black, Tressie Knowlton, Mrs. McKinney, and Nadine Knowlton.

A couple of the mothers usually went with us, or other neighbors, bringing with them delicious picnic treats like potato salad, deviled eggs, and roasted hot dogs. It was so much fun running down in the canyon, where everything was still green.

Our schoolhouse was more than just for school; it was used for other events as well. Christmas was a most eventful time. We practiced a play for a couple weeks to perform as part of the Christmas party with the whole community attending. For the stage, sheets for the curtains were hung from a wire. We had a Christmas tree and exchanged an inexpensive gift; we each got an orange and a few nuts. The school was the community center, where voting was done on Election Day, dances were held, and neighbors came for social events, such as box or pie suppers. Our rows of desks would be lined up around the edges of the room. Dances were for everyone: young, small, old, and all ages in between. Some of the men waxed the wooden schoolroom floor with a dry powdered wax, so that it was as slick and smooth as glass. I was thrilled when one of the older men would invite me out on the floor for a waltz or fox trot, and grateful for all the practice Mrs. McKinney had given us during school. The square dances were the biggest hit, with Mrs. McKinney on the piano, Paul Huntley with his fiddle, and someone else doing the calling. We'd be in sets of all ages on the floor for the "allemande lefts" and "swing your partners". For the pie or box suppers, the ladies would do all the baking or fixing and put it in a colorful, anonymous box. The men then did the bidding for each box the auctioneer held up. Naturally, they wanted to buy the box of their girlfriend or other special person they wanted to eat with for the evening. The women were not supposed to give out any hints, so there were some lucky men and boys, and some disappointments.

The Annual Wish Book

Near the end of each year, Montgomery Ward's large, thick catalog arrived, and our old one was taken to the outhouse to use for next year's toilet paper. Everyone knew to be conservative, as it would need to last an entire year. Some families received the Sears and Roebuck catalog instead. The mail order catalog was very important for ordering a few clothes, such as our underwear and socks, or new winter rubber boots, plus some necessary ranch items like milk strainers and filters or a lacking tool when transportation to town was time consuming and difficult. When I went to school, Mother ordered me two pairs of bib overalls and two shirts for the year, to be worn a week each and taken off as soon as I got home. The majority of items in the catalog were not available in local stores. No doubt it was a "wish book" for grown-ups eyeing all the things they needed or would like to have but did not have the money to buy. It was my dream book. I spent hours poring over the pages of western gear, saddles, bridles, and boots. Over the years, I did receive for a birthday or a Christmas needed boots and a leather jacket to last a life-time. But there was the never ending dream of a new saddle or bridle, but of course the old ones would have to do.

The Horses I've Ridden and Trained

There were a few of Mother's horses left on the ranch when Mother married Dad. Her horses had been running with the Sanford horses in one of the wild herds while Mother and Dad were in California. Bill brought them back: Babbie and two of her previous colts, Gypsy and a gelding. Only Babbie was broken to ride, and Dad rode her sometimes when he herded the sheep. Once when I was four and saw him coming home, I ran across the arroyo to meet him. He put me on Babbie's back, intending to walk along beside me, but she jerked away from him, bolted, and ran bucking for the corral, throwing me off. Mother, furious as a broody hen whose chick has been attacked, flew verbally into Dad, reprimanding him for putting me on an unsafe horse. Fortunately, I was neither hurt nor frightened by the incident and wished for another ride.

Babbie was our brood mare, who had a colt almost every year. Her first colt after we were back at the ranch became a huge gelding we called Bill. Bill Henderson tried to help us use him for plowing, but that was pretty hopeless, since the horse didn't know much and neither did we. Chief was next, a handsome horse with a lot of spirit, so much so that the man who was going to break him found him too much for him, so he brought him back. Jeff and Dad said he needed to be tied up for a long time to calm him down. It was hot August weather, and he got too hot on his back and died from kidney failure. Mother always felt bad about that, as tied in a hot corral the poor thing couldn't get out of the beating sun to any shade. I remember helping Mother carry water for him, so at least he'd have ample water to drink.

Ginger was the mare that trained me how to ride and work cattle when I was eight. She was a real cow horse, but between twelve and sixteen years old when Dad got her, she was considered past her prime. Nevertheless, she got the work done, and all I needed to do was observe and learn while clinging tightly to the saddle horn. She would jump and turn so quickly if a cow tried to get away that if I had not hung on tightly, I would have gone flying. I fell off sometimes, but not too often. Once I got better at riding, I started to ride Ginger bareback and learned to stay on with all her whirls and twirls. When cattle were to be put through a gate, we'd gallop ahead to open the gate, and she'd stop on all fours while I jumped off to open it. When getting back on, she wasted not a second to whirl around and was off at a gallop to get back to the cattle before I was fully settled on her back. She was a superb cattle horse, and did a lot of work the years that we had her.

Mother said if I was old enough to ride, I was old enough to take care of my horse. Ginger was to be curried before the saddle went on, being sure there were no burs to poke her back. In the fall, the cocklebours needed to be pulled from her mane and tail. She was not to be "winded" or ridden in unsafe places, to be offered a drink before I got mine, and when finished riding, the horse was to be unsaddled, put away, and fed before I could come in the house.

After Chief, Babbie foaled Laddie. A man that lived over by the Red Top said he'd break him, but he decided Laddie was more than he could handle. Laddie had run off a time or two and was gone for about a year. I was twelve when he came back, and I said to Dad I wanted to break him. Mother was extremely opposed, afraid that I'd get hurt or killed, as he wasn't broke, not even his spirit. Somehow I evaded her orders, and Dad helped me saddle him. At first, he'd walk and lead the horse. Later, as we still had my old Ginger, Dad would ride her and go out with me riding on Laddie. Gradually, by working with Laddie every day or evening when I got home from school, he got so I could ride him out alone. Before long, I was riding him bareback, although after he was broke he still tossed me bareback, once four times in a row. I got on, he tossed me; I swung back on, and again he bucked me off, until the fourth time I swung up he allowed me to stay.

Laddie was exceptional.

Laddie was an exceptional horse. Although we had up to eight or more saddle horses at a time: Charlie, Paint, Black Bess, Ginger, Dixie, Ginger II, a bay stray horse that was with us over a year, the stallion, plus the young ones I was training, all these horses put together couldn't equal the alertness, speed, and endurance of Laddie. Laddie became the horse of my life, the one I always rode when there was anything hard or important to be done. He mostly ate only the prairie grasses, but he wasn't thin, just lean, tough, and fast. He could fly for two miles or more and be terribly hard to pull up, as if he could run forever without becoming winded or puffing. I did ride him with a saddle at times, especially on the long rides when cutting cattle and finding strays, but we usually always went bareback. He was tall, and I'm short with short legs, but I managed to swing up by holding his mane and swinging one heel over his backbone and scrambling on up. If there was a gully bank, or something to stand on, it made getting on easier. He was not a horse to stand patiently around and wait for one to get on. Often he was sailing off at a full gallop before I was seated upright on his back. This was typical at a gate, if we were opening it to get cattle through. He seemed to know they'd scatter and cause a lot more work if we didn't get back to them quickly.

I rode Laddie for nine years before leaving the ranch. Over those days, months and years we shared several thousand miles in cattle work and pleasure. I trained and rode another dozen horses, some fine, others ordinary, and a few plugs, but one or all, none could 'hold a candle' to that plain-looking, sixteen hands tall, lean, brown gelding. I adored him as though he were some handsome royal steed, talked to him as a bosom friend, hugged him, and buried my face deep in the soft spot of his neck to soothe the sobs of a crushed heart. He substituted as the brother, sister, or playmate that I lacked in the isolation of my childhood.

Laddie's stamina was such that he could put in a grueling day working cattle and still be eager for an evening race into the wind. In Colorado there was bound to be a strong wind a blowin'. Often I rode him bareback with just a twine around his lower jaw. He knew my intentions so well, he sensed mental commands before the signal, a press of the knees or tilt of the body, was even given. Of all my pleasures, this one was supreme: to lean against the wind as Laddie charged across the prairie. I would spread my legs out from his sides, balance lightly on his back, lean forward, arms outstretched, palms

We "flew", sailing with the wind.

faced against the wind. We not only "flew" but we sailed with the wind, his long tail streaming out behind. The feeling of utter freedom and exhilaration, suspended between the power of the wind against my face and chest and the inertia of his speed, is unforgettable.

Laddie and I knew the prairie for miles in every direction, had our favorite canyons, and followed down the long cedar and pinyon clad ridges; we knew where the water springs were and where the various wild animals lived and raised their young. Sometimes we stopped to hunt arrowheads, to catch a little rabbit to bring home inside my shirt, to kill a rattlesnake with the knot of my lariat, to pick Mother a bouquet of prairie flowers, or perhaps to tease a silly old badger while he put on his fiercest temper act. In the fall, on the weekends, I rode slowly enjoying favorite haunts, absorbing the warm sunshine and that feeling of "fall is in the air today." In the winter, I enjoyed long rides when the snow came floating softly and quietly down. It was so beautiful, I felt at peace and perfectly content.

After Laddie, Babbie foaled a filly, Ginger II, who had a calm disposition and whom I easily trained. She was eventually sold for a child's horse. Next was Dixie, a pretty, shiny black mare with a blaze face and white stocking feet. A young man said he'd charge forty dollars to break her for us. Apparently, she was a hard one like Laddie had been. She came back broke, but the man must have struck her over the head. If one moved a hand quickly to catch one's hat, or any sudden jerk, she'd come unglued and gave me several bad spills. Once I managed to stay with her bucking panic for nearly a quarter mile before being thrown into shale rock on a ridge and getting pretty skinned up. I used her a lot at lambing time, often having to carry lambs on her, but I remained very cautious about quick movements lest lambs and I both get thrown.

Charlie, a younger brother to Laddie, was Babbie's last living colt. Charlie had a lot of spunk and was not easy to break, but I trained him entirely on my own, not wanting to have any more bad experiences from outside trainers, as had happened first to Chief, then Laddie, and later Dixie. I vowed we would never send away another horse to be broke. Charlie was a big brown horse and quite a bucker. Even after he was broke, every time when saddled, and before I got on, he would put on a display of bucking. That out of his system, he'd give a good day's work. Sometimes he would get clever and puff out his belly when being cinched to make the saddle loose when ridden. In time, he became nearly as proficient as Laddie. I think he prided himself on how well he could cut cattle, loving to outwit a cow at every turn.

Paint was a pretty brown and white pinto we paid forty dollars for. I really liked to ride him, and aside from Laddie, I rode him the most. I could do tricks on him, and tap him lightly in the flanks to get him to buck a bit for show. He could rear up nice and high; I loved that, and often had him rear up. If riding in sight of Highway 10 and a car happened to be going by, I'd have him perform. They most likely thought it was for real. He was a well-trained and fast horse; he never dumped me and was a joy to ride. He was a good one for cow work that took a lot of cutting and keeping renegades in the herd, as he'd dart out after any steer that took a notion to bolt.

Once, when I was driving cattle in Cactus Flat with Tom Hughes, the cattle weren't going too well, as we had to cut out a bunch from the herd to take away. Naturally, they had every intention of getting away from us and running back to the cattle they were with. I was riding bareback on Paint, and he was doing a marvelous job of out-dodging the stock. Tom Hughes, an older rancher and life-long cattleman, kept saying he'd never seen a better bareback rider. I must say, I did do well, and had not the slightest fear of losing my balance and falling off into the thick growing cactus that graced Cactus Flat. Those days were the prime of my life. Few people are blessed with as marvelous a childhood as I had, with so much freedom and plenty of adventures.

Years came when not only Paint but a couple other of our horses (fortunately, never Laddie, Black Bess, Charlie, or Dixie) got selenium poisoning. It was a terrible thing. Apparently the soil along the limestone ridges contains considerable amounts of selenium, which was taken up into certain plants such as Prince's Pine. When the horses ate these plants, they became overdosed with selenium, and its poisonous effects caused the hair of the tail and mane to fall out, and worse yet, their hooves came off, leaving very tender feet they could barely hobble on. It took a long time each time they were poisoned for the new hooves to grow out and cover their feet so they could walk with comfort again. Naturally, the horse could not be ridden for several months. After the third sad occurrence with Paint, I decided it was time to sell him, hopefully to a family in an area where this poisoning did not occur.

Black Bess was eighteen months old when Bill Henderson sold her to me for twenty dollars. He said he would have given her to me, but that he had bad luck when he gave horses to someone, so that's why he sold her. She was from the same mare as his favorite horse, Smoky. I loved her, but she did not have the sheer speed and stamina of Laddie, so even though I rode her often, she was not my main horse. I started to train her right away to lead and leaned my weight on her back. By the time she was two, I could ride her bareback. She was nothing like Babbie's colts, who each fought to the bitter end when being broke. She never even bucked. I couldn't believe a horse that didn't try to throw the rider at first. Soon I was working with her on reining, the gaits, turns, and stops. After I left the ranch, Mother said Black Bess didn't let any man ride her. Curly Krieder was going to ride her to take cake to the cattle for Mother when she had a broken hip, but Black Bess tossed him, so he used her as the pack horse instead.

A few years before I left the ranch, Black Bess, Dixie, and Ginger each had colts. For two years we had three colts a year, and I was in my glory riding all these young horses bareback in the corral until I had them broke enough to use a saddle and ride out on the range. First I saddled them in the corral and hung gunny sacks of rocks and rattly cans on the saddle to get them used to strange things and noises on their backs. After I got married, my folks sold all of our horses except for Laddie, who later died on the ranch, and Black Bess.

A young filly I'm beginning to train.

Mother brought Black Bess with her to California when she moved there in 1964 (see Epilogue) for her grandchildren (Tamara, Roxana, and Keven) to ride. Black Bess became attached to me on the trip from Colorado. She liked to lick my body as if I were her colt.

Dust Devils

The whirlwinds, or dust devils, common on our part of the prairie, captured my attention—mainly while I was herding sheep. I would watch one until it spun itself out of sight. They were nearly an everyday occurrence. Some were small and daintily skipped over the ground while spiraling around. Others were big, dark and powerful, enough to take down a homestead cabin. This was the fate many prairie cabins faced. They were obliterated so completely that scarcely a board of them remained to be found. One of my interests was to investigate abandoned homestead cabins before they were destroyed by the whirlwinds.

Mirages and Other Natural Phenomena

One frosty, sub-zero early morning, when I was around eight, I was on my way to town with my dad. This was in the nineteen thirties, when some old, vacated homestead cabins were still in existence, scattered here and there over the prairie. This morning one of these cabins was actually in view above the horizon, slightly above the ground, and not too far down the road in front of us. It would normally be at least five miles farther before it could even be seen from the road. "Look, Daddy, look."

My dad was baffled. I was astonished, not believing my eyes. How in the world could the cabin be there when we knew it wasn't there?

I wanted Dad to explain to me how it got there. "Danged if I know; must be some kind of mirage." But this was not a normal mirage as I knew them on very hot days, looking like water on the ground just ahead. Then when you got there the water was still farther ahead!

This puzzled me for years, and I have never forgotten that perplexing but real marvel of nature. As an adult, I learned how it is caused. What Dad and I had seen was what is called a "superior mirage", one that is seen elevated and apparently floating above the horizon. It also appears to be much closer than it really is. This can occur when the air on the ground is much colder than the air above, causing light rays to be bent down, producing a displaced image of a distant object to be seen above the observer's line of sight. This kind of mirage is more common in the Polar Regions. I don't know how cold it was that morning, but it was likely more than twenty degrees below zero. Rarely, the temperature in winter could get as low as thirty degrees below zero, but never again did we see a superior mirage.

Seeing a regular mirage produced by heat on summer days was very common. There would be "water" on the road ahead of one or a tempting "lake" of water when the land was parched. Sometimes at night during hot summer weather by the hydrant under the cottonwood tree there would be tiny, sparkly, greenish lights on the ground. They were mystifying to me. Mother explained that they were phosphorescent lights caused by bits of oxidized wood. In winter, when the temperature was very cold and the weather about to change, a storm brewing, we would see sun dogs (they're colored like a piece of a rainbow) on either side of the sun.

I have no clue as to how Mother knew we could see the northern lights, or aurora borealis. We didn't get a newspaper or have a telephone. It may have been before we even had a radio, as I was fairly young. Anyway, she knew they would be visible and that we couldn't see them from the house, as it is situated at a lower elevation than was needed to see them. It was a cold winter night, but Mother and I bundled up and took a mile-long, nighttime hike to Haystack Butte, which we climbed to the top by lantern light. It is the highest point around. She had us look a little west of north toward Pueblo, telling me that was the direction to see the aurora borealis. On the outskirts of Pueblo is one of the largest steel mills in the United States, which works around the clock. Mother had feared that the glow of the steel mill, some forty miles away by crow flight, might affect seeing the northern lights. She needn't have worried, as the faded orange glow of the mill penetrating just above the horizon line was no match for the brilliance of the northern lights.

We stood and watched, waited still longer, and kept looking. Not moving very much, we were getting quite chilled, but Mother kept my hope high for about an hour. Then it happened! We saw the northern lights shooting in the sky like curtains of light! It was probably not as spectacular a display as if we had lived in Alaska or far northern parts of the world, but we actually got to see the aurora borealis shimmering brightly above the horizon. The effect, caused by solar winds blowing across magnetic fields and colliding with air molecules near the North Pole, must have been extraordinarily powerful that night for us to be able to see them at our latitude. In later years, I read that every eleventh year they are extra spectacular and can be seen at lower latitudes. This must have been one of those years, and somehow Mother knew this was the right night. She had created one of those impressive moments of my childhood.

On a special night each year in mid-August, Mother and I took a blanket to sleep outside and look up watching all the "shooting stars". It was exciting to see so many, as I kept pointing and exclaiming, "Look!" "Look at that big one!" "Oh, see that really bright one over there!" Normally we only saw a shooting star once in a while, and Mother said one could make a wish. Another fun thing was to see who could find the first evening star to appear in the sky and then say: "Starlight, star bright, first star I see this night. Wish I may, wish I might, have the wish I wish this night."

With the ranch elevation at 5000 feet and a nearly always clear sky, nights were jam-packed with stars. Mother and Dad wanted me to be able to always recognize the Big and Little Dippers in order to find my directions in case I got lost. The brightest and last star in the tail of the Little Dipper is the North Star, always above the north pole. It has been used since ancient times to tell direction. The top, right star in the cup of the Big Dipper always points to the North Star.

Sounds

There were a myriad sounds on the ranch, such as rain falling on the roof, hail banging on the tin barn roof, or tremendously loud claps of thunder, down to the gentler thunder referred to as the rumble of the potato wagon. The wind had all kinds of sounds. It could whine, sigh, and roar. It could be gentle, causing quiet laps of the water on the edge of ponds or a whistle through the tree leaves.

The animals had their sounds: Horses their munching of hay, stomps, whinnies, and snorts. The cattle bawled, mooed to their calves, or banged against the walls. Our chickens crowed, cackled, sang, clucked, and peeped. There was seldom a perfectly silent moment.

There were two sounds that baffled me. Both occurred when the temperature ranged below zero. The Santa Fe train ran some fifteen to twenty miles southeast of our ranch. We never heard the train except on a few rare occasions when it was extremely cold, when it sounded so near that it seemed only a quarter mile away as it rumbled down the track and blew its whistle at a cross road. The other sound came four miles across the prairie in the early evening from just below the crest of one of the Sanford hills. This was where our closest neighbor, Bill Henderson, lived. When he chopped his wood on extremely cold evenings, we could hear the sound of his chopping as clearly as if he were splitting wood by our place. Otherwise, we never heard him. Strangely, we could actually see him chopping the wood, but the sound of his chopping reached us seconds later.

When in canyons I liked to hear my yell repeated in echoes against the sheer rock walls. I'd make different sounds just to hear the echoes come back.

Weather and Storms

Nowadays most people check the weather forecast by computer, television, radio, or newspaper, rarely an almanac. We did have an almanac. Knowing what to expect ahead of time from the weather is important to ranchers to be as prepared as possible to care for their livestock and garden. Mother did her best to predict the weather by interpreting the many natural signs.

One sign was a ring around the moon predicting moisture coming soon. Another was the direction of the wind and way the smoke blew from the chimney. In summer, how the thunderheads came over the Rocky Mountains, the time of day, location, and how dark a color they turned as they headed our way, or missed, were eminent signs. When they turned extremely dark, all but black, it was a pretty sure sign of a hail storm or at least a bad thunderstorm. It was time to get the animals under shelter, and quickly if the distant roaring sound of hail was heard. If I was out riding, I quickly raced for home to get in the barn before it hit. Twice my horse and I got pelted by hail just before we made it to the barn. Animals often predicted a storm, especially the horses. They'd become very antsy and race in big circles in the pasture kicking up their heels, energized by the higher oxygen or ozone content in the air. The sound of distant thunder was a sign a storm was coming, but not necessarily one hitting our land, as we would stand smelling rain in the air (likely one of a rancher's favorite smells—certainly one of mine) while watching the clouds drop their load elsewhere as they passed on over.

Thunderstorms were very spectacular with loud claps of thunder and dazzling flashes of lightning. Besides the arroyo flooding during such storms, water gushed down the slope above the house and ran into the house as much as four inches deep on its rush down to the arroyo, entailing a muddy-silt mess to clean up. We hastily tried to get things off the floor before they became soaked. After the rain, the sun usually came out shining brightly, and to the east a brilliant rainbow would appear, frequently with a second one above the first.

I'd heard of the pot of gold at the foot of the rainbow, and a number of times I went to search at the foot of a rainbow that was not far from the house. Each time I was almost to the spot where I saw it ending on the ground, it would be farther on, forever alluding me.

A cold wind "out of the east" was a probable sign for snow in winter. When the snow was deep, we walked single file, the first person breaking the trail. If the sun was out, the frozen snow just glistened with fiery sparkles. When the snow lay on the ground, the livestock could not graze, so we prayed for a Chinook Wind, which seemed to quickly melt the snow into the ground. Then we'd slosh in the prairie mud, often having a hard time to pull one rubber-booted foot, sunk in the mud, out after the other with the suction of the mud plus the heavy mud balls that clung to our boots.

After a siege of stormy weather, if Mother saw a break in the clouds with some blue sky, "enough to make a Dutchman a pair of breeches", she'd say, the prediction was for fair weather.

The Hail Storm from Hell

There had been a terrible hail storm late in the summer with small stones bigger than marbles, many the size of bantam eggs, and even a few like hen's eggs. The big ones were enough to kill an animal. The storm came on quickly with a very black cloud in the west, but soon there was a terrible roaring noise. Mother, who came from Iowa, thought it must be a tornado and sent me into the cellar and threatened me "NEVER STICK YOUR HEAD OUT!" She and Dad worked as fast as possible driving the poultry into the barn. I remember them recounting how stupid the turkeys had been, just standing there with their heads in the air. My parents grabbed them and threw them into the barn to get them under protection before the fast approaching storm struck. That was all the time they had before the deluge of hailstones struck, and they dashed into the house, leaving me in the cellar. The hail storm was accompanied by a torrential thunderstorm, and the water that raced down the hill and road washed the hail stones ahead of it and into high drifts. When the pelting of the hail stones ended and Mother let me come out of the cellar, I was amazed to see the big piles of hail stones, some a foot and a half high. Most of the panes in the windows of Bertha's homestead were broken, the watermelons in the garden, just ready to eat, were demolished, and the rest of the garden as well. The few turkeys they missed catching were killed, and some sheep and cows were hurt but not killed. The water gaps in the arroyos were all washed out from a higher than usual flood. That was one hail storm that will never leave my memory.

Mother was working around the ranch getting things back together the next day. That forenoon Dad set out on foot to survey the damage to the range and find where the sheep could graze. I went along. It was a sad sight. The grass that had existed, which wasn't much, had been chopped off by the hail and then the flooding caused by the rain that accompanied the hail had washed it away, leaving the ground quite bare.

We hiked as far as Cactus Flat. That's where the folks chopped the cactus to make the ensilage they fed to the milk cows. It looked like a deserted desert with sticks and stalks standing up out of the ground. The hail had beat the branches off the cactus. Dad wondered if the cactus had been killed or if they would grow back. With their cactus tree protection gone, jackrabbits that had taken shelter under the cactus were knocked out or killed. In those years, the jacks were still very prolific. Now we saw many dead jacks stretched out here and there with water soaked fur. I suppose the vultures had easy meals for some weeks to come after that storm. We hiked some three more miles to Averson Canyon, appalled by the damage along the way. We even saw a dead civet cat, a rare sight, as they are a very secretive animal. Being quite small at the time, this was the longest walk I had ever taken.

When the folks next took the sheep out to graze, I remember them coming home bleating with hunger, as they could not find enough grass to eat. There really was nothing the folks could do for them but keep taking them to a different place on the ranch each day until the grass started to grow quickly from the rain and hot days. Then at last the sheep finally got a stomach full.

Rattlesnake Encounters

Since there was a rattlesnake den on our ranch, which we had not the slightest idea existed, rattlesnakes were always numerous around our place. The year I was three my parents killed sixty rattlers. Needless to say, my mother was always nervous that I might get struck while speeding around everywhere. For many years our average rattlesnake kill on the ranch remained at thirty or above.

I started herding sheep on my own at age five to six, keeping the small flock fairly close to home so the folks could keep an eye on me at first. Walking around herding sheep is an excellent way to find rattlesnakes as well as many other interesting prairie creatures. It is not so nowadays, but back then it was supposed to be worth a day's work to kill every rattlesnake found as an insurance against having the livestock snake bit. Ranchers needed every dollar they could make and just couldn't afford to lose any stock. If I was in screaming distance of the house, one of the folks would come kill the snake, but if not, I would try on my own if I could find rocks to throw at its head or a large enough cedar stick. These earliest years I didn't have a pocket knife or think of saving a snake's rattles for a trophy. Later, I did so, and most especially on the big fellers with lots of rattles, so I could brag about how big the snake was and what a battle it had been to kill it.

From about eight on I had a horse and carried a pocket knife and lariat so my collection began. The knot at the end of the lariat was perfect to swing down hard on a snake's head, providing I could hit it on the head! When I was sure that enough blows had properly landed to kill the rattler, I always held it down with my boot heel to cut its head off first for safety before attempting to take its rattles.

When I was about nine, I was riding Ginger some four miles from home in the Red Top pasture. I came upon a mother rattlesnake, and on dismounting to see if there was any way to kill her, I not only realized that I had forgotten to take my lariat, but far more amazing, discovered that she had twelve of the cutest babies about the length of a lead pencil, each with a teeny button at the end of its tail. Never before had I seen a mother rattler with babies. The folks had said it was worth a day's work to kill a rattlesnake for the safety of the livestock. I knew she would have to be killed but there were no rocks or strong sticks nearby. That meant a trip home to get a shovel. Seeing no landmarks to identify where to find them again, I unsaddled, left the saddle by the mother snake, and rode home bareback to get a shovel.

At home I told the folks about the babies and that I was taking a shovel back to kill them. Dad said, "You're not going alone!" He was not able to ride bareback, so he walked and I rode. When we got near the place where I had left the saddle, I hurried on ahead to make sure they were still there. When Dad arrived, he killed them all, thirteen rattlesnakes at one time. When the local newspaper heard about this, they put a small article in the paper—my first write-up.

One huge rattlesnake I never got, but it came close to getting me. In some places the ground erodes, leaving holes about eighteen inches deep and tunneled back for a ways. In one of these places I spied a cute little baby jackrabbit. I loved little bunnies, as all children do, and I wanted to catch it and carry it home inside my shirt for a pet. I got down on my knees, then on my stomach to reach a hand down into the hole to catch it. But it was strange that the little tyke didn't seem to notice me and sat quivering, looking straight ahead. On my belly, I inched closer quietly to grab it, and there not eight inches in front of the rabbit was a huge rattler. I was so startled I backed quickly away. Luckily, the snake remained intent on its rabbit prey and didn't turn its attention to me. Otherwise, I could have been struck in the face when I had my head in the hole.

When I was about ten, Jeffery's family got sheep and my folks ran our sheep with theirs and a sheepherder was hired. But I still herded small bunches at lambing time. I was tending a flock near some buttes on top of which grew a few bushes we called white sage, usually about twenty-four to thirty inches high. The sheepherder had the big band up that way a day or so before, so when I found a "dead" rattlesnake hanging on top of a sage bush, I picked it up and put it on the ground to cut its rattles off for my collection. That sure brought the snake to life in a hurry, and me as well! I killed it properly before again attempting to remove its rattles. I had previously never seen rattlesnakes sunning on top of a bush, always on the ground, so naturally I thought the sheepherder had killed it and then tossed it up on the bush. Crazily, I made the same mistake another time when trying to cut the rattles off a very dead-looking snake on a bush. It was definitely not dead—a shocking experience for both the snake and myself.

Laddie didn't like the smell of rattlesnakes. When I'd get off to kill a snake, I couldn't very well hang on to him, and many times he bolted and ran off, reins flying in the air, leaving me with a long walk home. Rattlesnakes were often found in the shade of a fence post or cactus bush, and if I was lucky, I could tie Laddie first, but that was rare.

One time I found a rattlesnake near a cedar post lying on the ground where a water gap had washed out. As I started to hit it with the post, another snake rattled right behind me. That sent a shock right through my chest as I jumped sideways to safety. They were a large pair, very protective of each other, and very hostile. They kept moving toward me and lunging out at me with their full length and then some. It took me a long time to kill those two, enough to give me nightmares, but I finally got their rattles to add to my collection. Two other times I came upon two at once, and each time I only noticed one until the second came up from behind, giving me a great scare. They seem to be extra fierce fighters when in pairs. I never killed less than ten rattlesnakes a year while living on the ranch, so the rattles I collected accumulated over the years.

SIXTEEN: "Dear God, I'm too young to die!"

My husband was shaking me saying, "Wake up, Barbara, wake up, you're screaming. What's the matter?" Trembling from my harrowing dream, I stammered, "A huge rattlesnake just sank its fangs into my leg. Two huge rattlers were coming at me fast. I tried to run from them, but I could barely move my legs, and then they became so paralyzed I couldn't budge either leg. The feeling of not being able to escape was terrifying!"

Unable to fall back asleep, I lay remembering that night of horror when I was sixteen, which has caused nightmares such as this to recur. That night is now decades ago.

It began as a typical Friday after school, with getting the saddle horses foremost in my mind. No school bus ran in our remote area of southeastern Colorado, making it necessary for me to stay in town with a family during school days, as the high school was thirty-five miles away from our ranch. When school let out on Fridays, Dad was there waiting with our blue Chevy pickup that he always let me drive. Eager to get home, I pushed my foot impatiently on the gas pedal despite the many chuck holes on the way. As we headed down the dirt road, the streak of dust behind us blew eastward with the wind. The autumn sun was just slipping over the horizon as we pulled in. "We're home. I'm going to find the horses," I called to Mother, who was in the garden, before dashing into the bedroom to slip on old jeans and boots. Fresh-baked cinnamon rolls were on the table. Mother was always thoughtful. I grabbed three rolls, hurried to the barn for a bridle, and was off on foot to hunt the horses as I ate.

Which way to look first? The horses could be most anywhere on our four thousand acres. "I know," I thought, "climb Haystack Butte," a little over a mile from the house. From the top one could see the ranch in all directions. Soon I was scrambling up the steep side as fragmented limestone slipped underfoot. Vegetation was scant on the butte, mostly woody sagebrush and scattered clumps of grama grass. I met one small rattlesnake about half way up the butte but kept climbing. Walking along the level top, my eyes absorbed the expansive view past our ranch for fifty miles (as the crow flies) to the Spanish Peaks, Fisher's Peak, and Greenhorn Mountain, landmarks of the Rocky Mountains. Suddenly a rattlesnake gave a loud buzz and instinctively I jumped and began to watch my steps. Amazingly and most unusual, I passed five more very large rattlesnakes before reaching the east end of the butte. "How come they're up here?" I wondered. I'd climbed this hill so many times since childhood, and though rattlers were frequently found near the foot of the butte, this was the first time I'd encountered any on top.

Standing at the east rim I scanned in each direction. The hope of spotting the horses blotted the snakes momentarily from my mind as my eyes searched the cactus-studded prairie, green draws along arroyos, and the cedar ridges. Were there horses among the trees? Did I discern a movement? With twilight beginning, shapes were becoming deceiving. "Oh well, it's getting too dark. Tomorrow at dawn I'll check out the south pasture; they must be that way," I thought. How beautiful this country looked to one who had been in town all week. I stood absorbing its magnificent views, bridle in hand, gazing down the range of Sanford hills, those cedar dotted limestone bluffs that took up most of the east; turned slowly north to view the Red Top ranch, their bunches of grazing white faced cattle mere moving dots; and then faced westward where stretched all those intriguing, luring canyons: Lone Jack, Meyers, Plum, the Cross Bar.

For years I had longed to ride over and explore them, but Mother was emphatic that I shouldn't go alone. Besides, we were always too busy to go. "You know," I pledged to myself, "I'm going to ride down every one of those beckoning canyons before I die." My thoughts shifted to imagining I was riding down Lone Jack with the slim, disdainful cowboy I worshiped at high school. He never paid me the slightest attention, but now he rode by my side as we talked and exclaimed over the sheer rock walls and spectacular scenes at every bend of the canyon. Minutes had slipped by before the realization that it was getting quite dark jolted my senses. "Why didn't I stop my silly daydreaming sooner? I gotta hurry now!"

Before taking six steps a rattlesnake gave a loud, threatening warning, its gray blotchy color blending so well with the shale I could barely see it. "Oh dear, I should have remembered those big snakes and left much earlier while I could still see them. There's no moon tonight either. Why did I ever let it get so dark? Well, I'll just have to take it slow and be real careful." I stepped gingerly to the right, but there an even meaner sounding rattler violently buzzed. A frightening feeling shot through my chest. I stepped back slightly, only to have both snakes fiercely rattle at the same time. I froze, petrified with fright. "What ever shall I do?" I could picture vividly those two barely-visible shapes: cold black eyes fixed on me, bodies coiled, heads arched and aimed, necks S-shaped for power, poised for the strike. My instinct clamored desperately to bolt in panic, but my intellect insisted, "No, you fool, you can't." I stood like a statue with a thumping heart between the two, praying their strikes to be stayed. "Oh God, please save me!" Eventually their ominous steady buzz

Surrounded by rattlesnakes when nearly dark.

ceased and they moved on. But with the dusk many more snakes were slithering around everywhere. Would I be able to stand perfectly still if one crawled over my boots? I could hear soft, light shuffling sounds on the shale. Reality struck cold into my heart: there were not just six rattlesnakes up here with me; I was now surrounded by dozens! I found myself praying fervently, "Dear God, I'm only sixteen. There's so much I'd like to do, and all those canyons I've never yet been down. I'm too young to die now."

I was certain that, even standing perfectly still, enough rattlesnakes would strike, sinking their fangs into my legs, possibly even through my boots, injecting more venom than I could survive. "Please God, help me, and I'll try to make it up to you somehow. I'm just too young to die." After several more prayers to God begging, pleading, and promising, a psalm passed through my mind: "Yea, though I walk through the valley of the shadow of death, I shall fear no evil, for thou art with me." I relaxed enough to inch my way along. Unable to see the ground by now, I had to depend solely on my ears. Every way I tried to move the nearby buzzing of other rattlesnakes struck my ears. Then I thought of a way to detect where the snakes were without seeing or stepping on one. I reached down slowly and quietly, sliding a hand down between my legs to the ground to pick up bits of shale. I tossed out a small piece with deliberate care so that it might land gently to inquire and hopefully not antagonize any snakes. If my pebble produced no answer from a rattler, I would move a little way in that direction. When I received a vehement response from a disturbed and angry rattlesnake or worse yet, several at once, I stood motionless for a while, silent except for the thudding of my heart, before tossing a bit of shale in another direction.

This ordeal went on and on, with me weaving in and out, back and forth, among the slithering rattlesnakes, for about three hours, which seemed like an eternity. Eventually, I had worked my way about twelve feet, but near enough to what appeared to be an edge of the butte, where I hoped to make my getaway. It happened to be the side opposite from home, but by then any edge was a tremendous relief.

"Now!" I thought, "Now is the moment." With all my might I leaped upward and out over the edge and bolted down the steep hillside, causing shale and rocks to tumble down after me. I ran over a couple rattlesnakes, perhaps more, their rattles furiously vibrating, left in the flying debris with absolutely no chance to strike with my downward speed. My feet were just marvelous in the dark: no tripping, no stumbling. I was down in minutes, flew around the base of the butte, and raced on for home in the night.

Within a quarter mile of home I saw a jerking, flickering light. Mother was coming up the road carrying a lantern. As soon as I got near, she lit right into me with a good scolding, "You know far better than to stay out way after dark like this without letting anyone know where you are. What, for heaven's sake, were you doing all this time? Where have you been? Why didn't you come home before it got dark? I've been worried sick thinking something must have happened to you. Don't you EVER do such a thing again!" By the time Mother had finished letting built-up concern off her chest, I'd caught my breath from the fastest mile I'd ever run. Mother would never have had the heart to give me such a scolding if I'd had a chance to tell her first about the horrible ordeal I'd just lived through.

The next morning, after I had the saddle horses in, my cousin Jeffery, also sixteen, came out from Fowler. I told him about my petrifying experience of last night, but his expression showed he thought my snake story was quite exaggerated. I suggested we ride up to Haystack Butte and I'd show him. Tying our horses at the base of the butte, we climbed up and walked along the top, not seeing even one snake. Jeffery didn't say a word, but I was greatly disappointed because now he'd never believe me. Almost at the end of the flat top, we suddenly saw a sight that halted us in our boots. We gasped in astonishment, and even after last night's experience, I could scarcely believe what I saw. Near a small, narrow crevice in the top of the hill, a big mound of rattlesnakes was piled up, one on top of another, in one huge entangled mass of all sizes. A conservative count would be at least sixty, but to us it looked like a hundred. Another dozen were lying nearby sunning.

"Boy, Jeffery, can you believe it, our Haystack Butte is a rattlesnake den!" I told him I had once read a story about a man getting trapped in a den of snakes, and had thought to myself that nothing could be more terrible than that. I never imagined it would happen to me.

Jeffery was still shaking his head in amazement, saying, "You sure didn't make up your story this morning, but it was hard to believe." He told me he had heard of some men who made a raid on the Rattlesnake Buttes twenty miles west of here and gotten over two hundred rattlers. We wondered how many more snakes were inside our butte in the den beneath the crevice.

We struck at a couple of the sunning rattlesnakes with the lariats we carried, but sensing danger the mass of snakes began to slither down the crevice and go inside the butte to their winter quarters. Soon none were left. Had they finished sunning before we arrived, I could never have convinced my cousin, others, or even known myself how many snakes I had been stuck among the night before. How had I walked in the dark through all those rattlesnakes and remained unharmed? Did God also think sixteen was too young to die?

Reptiles and Amphibians

Other than rattle and gopher snakes, which were prolific, there were some red racers, our longest and fastest snake. They were excellent at camouflage by getting in a yucca bush that had pink bud tips. The racer would hold its head and about fifteen inches of its body straight up looking just the yucca's pink budded stalk. They fooled us a time or two into giving a quick jump, as we were startled to find a snake in the yucca bush. There were hog nose snakes, which were venomous and quite short, barely eighteen inches long, looking like a small rattlesnake minus the rattles. We also had garter snakes in our garden that evidently dieted on bugs.

Long collared lizards lived around the canyon, their tail longer than their body, aptly named for the black collar around their neck. They were fascinating to watch, but moved too quickly to catch, except for once, in which case the lizard was brought back to California and kept in a terrarium. Fence lizards lived around the corral fences and were often seen clinging to window screens to catch the flies. Alligator lizards, with very long tails that broke off when caught, frequented our board piles. They would hiss and act scary. My favorite lizard of all was the little horned toad (actually a lizard), which could be seen sometimes, almost camouflaged, on a bit of bare ground among grama grass. At other times they were near ant hills. The young, about the size of a silver dime, were easy to catch and a temptation to take home.

Frogs, toads, and salamanders were the main amphibians on our ranch. The frogs would sit at the edge of the reservoirs or other water sources hidden in pond grasses and weeds. When they saw me sneaking up, they would leap into the water, swim underwater, and emerge at the other side. Ever alert, their big eyes located nearly on top of their head, they saw everything. We had many toads, some gigantic, around the garden, obviously grown up from the many tadpoles I had brought home and turned loose in the garden when they began to hop. Mother appreciated their help catching insects around the vegetables. We called salamanders mud dogs; they liked to stay in damp places. Ours were mostly plain black, but the more scary ones, about nine inches long, had big yellow spots. They were often in the cellar, and I was afraid of them when I was small.

Insect Life

Insect life was abundant on our ranch. Some of the dung beetles were truly magnificent, having a magenta crest and iridescent green body. Usually near cow paths, they would be busy rolling cow dung into round balls in which they deposited their eggs. I wanted to show them to my children when back at the ranch on visits, but, sadly I've not seen them again, just the more prolific brown dung beetles, which are not nearly as exciting. Tiger beetles, about an inch long and fairly slim with a red body and iridescent green on their head and thorax, flew into the house at night. They were attracted to our lamp's light. Assassin beetles had colorful red markings. Black stink beetles were sometimes near ant hills, often with ants clinging to their slim legs. They usually walked with the tail end tilted upward, perhaps ready to stink if necessary. Bessie beetles (a ground beetle) were black with two moderately broad oval sections closely connected in the middle. I'm not sure Mother knew the correct names for all the insects, so I'll just say that we had loads of June bugs, lady bugs, cucumber beetles, and both kinds of blister beetles, gray and stripped. True to their name, they produced huge blisters almost an inch long caused by a poison they secreted. Carrion beetles, olive/brown with black markings, appeared around dead things. Our water tanks and pasture water holes contained many kinds of water bugs.

The ant species most prevalent in our area was the red harvester ant. I was always getting bitten by them; their bite made an itchy lump about the size of the end of a finger. Mother made a soda paste to take away the itch, at least for the time being. We had termites that lived under damp wood. They grew wings when swarming, and the queen termites were extra large. Mother tried to eradicate them, since they could do potential damage to the house. In summer, we had millers by the thousands fluttering against the screen outside, and hundreds that got inside were attracted to the lamp light, sometimes getting inside the chimney and burned up. They were truly a plague.

The prairie was also home to inch or more tall, hairy legged, dark brown tarantulas. I observed them but was afraid to touch or be touched by one. Their home was an inch round hole bored straight down in the ground with a half inch of grass thatching shielding it above ground. The tarantulas looked so huge that I never figured how one could squeeze down its hole. They were most often seen in the fall or nice warm days ambling around the grass sod, never venturing far from their burrows.

When the milkweed was in bloom, the flowers attracted very large, orange and black milkweed wasps. There was a patch of milkweed to go through on the way to bring the milk cows home in the evening. A few times I got stung. Their sting is so painful, it feels like being burned with a red hot ember. They also liked the snow-on-the-mountain plants.

We had some mosquitoes, but it was the gnats that were most annoying when riding home near evening, especially close to the canyon. There would be swarms of them. They got in my ears, eyes, nose, and mouth if not kept shut. A bandanna was helpful to at least keep them out of one's ears. Oftentimes, a sudden and delightful vein of warm air came down a draw—delightful if not laden with gnats.

Most summers, locusts sang loudly in the tree cactus about the time the cactus bloomed with velvety magenta colored flowers, which I considered the most beautiful of flowers. I always looked forward to the summer locust chorus, which could be heard for miles. The locust song is not too unlike the rattlesnake buzz, especially if their wings get stuck and vibrate to get loose. Many are the times I mistook a locust's vibration for a rattler's buzz and made a startled jump for safety.

Once when I was three, Mother had a hunch to look in the bed with the lantern before I got in. That part of Colorado has very big centipedes. She found a seven incher, which had crawled in the bed between the sheets. I remember seeing other huge centipedes she found in the kitchen among towels or my clothes. Around the barnyard it was not uncommon for me to find three to five inch centipedes under boards and rocks.

Ranch Dogs

We only had one dog that I considered belonged to us. He was just a little thing called Lucy. He had short, orange-red, wiry hair. Lucy was a Mexican dog Dad got in Walsenburg from a Mexican family that had a sheep dog. He thought we needed a dog to help herd the sheep and save a lot of steps. Lucy was a male, and, as Mother explained, his Spanish name was probably Lucero or Luchito. He did know a lot about sheep and how to herd and work them. Just wave an arm with instructions, and as long as he could understand what was wanted, he would do it. English was a new language to him. He knew what to do; he just couldn't understand what was being said. I loved the little dog, who was with us for a year before our sheep were merged with Jeff's band, and Lucy went to live with the sheepherder. Mother baked Lucy dog food in cake pans, and cut it into squares like cornbread, which he loved.

Ranch Cats

We had many cats over the years. When I was eight, we had a big golden tom cat. I wrote my first poem to this cat, which reads as follows:

Golden Tommy, my lovely cat,
There's always sunshine where you're at.
The music of your joyous purr
Is soft as is your golden fur.

We had a tabby cat named Tabby. Tabby's kitten, Polly, became one of our most memorable cats. She was a blue Maltese, who had multiple litters of kittens. Her most unique characteristic was all the full grown jackrabbits she'd catch, drag home, and lay down at the front door. The bigger they were, the more proud she'd be of her catch, and she would meow loudly until we came out to praise her. She acted as if we were her kittens, and she had hunted to provide us with a big meal. We had a rock wall and a gate, and Mother and I were mystified by how Polly could carry a jackrabbit as big or bigger than she was and jump with it over either the rock wall or the gate. When Polly died, we buried her with a rock for her tombstone on which we inscribed "Polly did her part."

Cats were fun to have around at milking time, but they could also be a nuisance. It was entertaining to see them stand up on their hind legs to reach for the squirt of milk I'd squirt into their open mouth. It was all right for a time, until the cats became too demanding for the milk. It didn't go over well with Bossy, the milk cow, when a kitten got excited about the milk and started climbing up her hind leg. Naturally, she jumped and kicked, and I was lucky to save the pail of milk. As long as the cats didn't scratch or touch the cow's legs, generally things went all right.

Woolly, My Hunting Partner

Woolly on the rock wall.

My best pal and loyal partner was Woolly, from one of Polly's later batches of kittens. He looked like a bobcat and was larger than any of our other cats, so we thought he might be part bobcat, or lynx, if they can crossbreed. Our cats were all short-haired, and this fellow had fur almost two inches deep. He also had tufts in his ears, not as pronounced as wildcat tufts, but much more so than in regular cats. He liked to sit on Mother's rock wall and in rocky areas that were natural to a wildcat. I called him Willie actually, but since he had woolly fur, he became known as Woolly. A typical cat, he was very demonstrative to the family, but became scarce if a stranger appeared. When I'd been out riding and came in very hot and tired, one of my favorite relaxations was to sit down and play the piano for a while. This practically had to be done when Woolly was outside, as he'd climb up and walk back and forth on the piano keys so as to get all the attention. He also liked to get in front of any book that was being read, but that wasn't as annoying as not being able to play the piano and having to listen to his purr music instead. He was persistent and

Woolly by Mother's rock wall.

not easily discouraged. When I went away to college, he seemed to know I was leaving. I'd have a suitcase opened on the floor to pack, and he'd spend his time lying inside it on the clothes. He also became extra affectionate.

I didn't have a dog, but Woolly acted in every way like a dog would have been for me. We hunted together and hiked for miles down the arroyo. When on a walk for enjoyment, he seemed to know the difference and padded along, tail up straight, showing every sign of happiness, with no worry of how far we might go. He was wonderful for hunting, not only around the corrals but down in the arroyo or wherever we might find a wild animal to catch. We caught a lot of cottontails in the board piles around the barn, each of us going in at a different end to scare the animal out to the other one. Sometimes he made the catch, but more often I did, as he must of have appeared scarier to our prey. The same was true around the arroyo banks, where animals thought they had a good hiding place, but if the cat could get in and scare them, I'd grab them when they ran out. Sometimes I'd throw clods of dirt to scare an animal in Woolly's direction, and he would pounce on it. He was a gentleman, never seeming to mind if I carried our game home as he pranced proudly beside me, tail held high. I guess he figured he'd get some anyway when it was skinned, cleaned, and cooked. I don't recall him hanging around the other cats; he only demanded our attention.

During weekends home from high school during wintry weather, Woolly and I would walk a mile or more in the snow to a bend in the arroyo east of our place, near the Sanford fence, to track and hunt rabbits. I always felt proud to go on an outing with Woolly, and he seemed proud as well, as seen in his stance and demeanor.

Jackrabbits and Cottontails

There was an abundance of jackrabbits in the early years—jackrabbits by the hundreds if not the thousands. Some ranches had jackrabbit drives to thin them out, where I heard they'd capture a few hundred of them. It must have been quite an exciting day for the men, but I felt sorry for the rabbits. When driving after dark, jacks were always popping up and dashing across or alongside the road. In each succeeding decade there were fewer and fewer of them, and nowadays when back at the ranch, it is not uncommon to not even see one all day. When I herded sheep, you could almost always see one loping someplace, or if startled, running off with long, high leaps every few yards. The leaps were to see what the pursuer was doing and to know how to time its circle, since jacks never left their home territory very far, even though it looked like they were running clear out of the country. They were often in pairs, but where they bedded down in their shallow excavations under a clump of grass for protection, always facing outward, there was only one, never two side-by-side or even close by.

No matter how many I saw in a day, it never ceased being exciting to have a big jackrabbit suddenly jump out from its "form" practically hidden under a clump of grass or from behind some bushy weed, almost underfoot, and go leaping off in huge ten-foot bounds with an extra high leap every so often to get a view of the situation. Their long, seven-inch ears would be straight up to detect sounds of pursuit. They'd lope off only a short distance and then circle back as soon as one left. When on horseback and wishing to chase a jackrabbit for fun, the jack would lay its ears back over its shoulders, stay low to the ground, and run all-out while swerving back and forth to be confusing and hard to catch. They were too fast to catch even on horseback. In this situation, they still tended to run in a wide circle, heading back toward where they started off.

Jackrabbits were a natural part of the land. They made life interesting for me as a child. Adorable baby jackrabbits were born during spring and summer, with usually three to five in a litter. Unlike tame rabbits, they are born furred and with their eyes open. They also don't get a furred nest and a mom that stays with them. She finds each one to nurse at night, but I never saw a mother jackrabbit by any of her babies during the day. The little one's way of keeping safe is by keeping very still. Little jacks, tiny and sweet as they are, get weaned and abandoned by their mother when about a week old, left to eat grass and fend for themselves.

The cottontails were even cuter, and it was these that I hunted for mother to fry and serve with milk gravy to go over potatoes or toast, a never failing treat. The cottontails were seldom found on the open prairie, but were much more prolific around the barnyard board piles, along the arroyos, and near the canyon, where they'd run under the large outcropping sandstone slabs.

Coyotes

One evening event that never failed to thrill me was the twilight chorus of coyotes howling from all directions. It seemed as though there were dozens surrounding the ranch house and corrals. I knew there weren't, that they only made it sound that way due to their ability to project their forlorn howls. Yet some howls seemed so close and others farther away, as if they were perched high on every butte or knoll in the area. I would listen intently, trying to figure out where different coyotes might be stationed as they sent out their long-drawn-out howls of varying pitches and answered back and forth to one another.

Like the jackrabbits, the coyotes were indigenous to the environment. I scarcely missed a day without seeing at least one. The largest number I ever saw in one pack was twelve. It was late one fall afternoon, perhaps around four o'clock, out in the open not too far from a cedar ridge in the Woods pasture. I eyed them, amazed at so many, but rode on about my business. The coyotes stood to look me over also, then continued to trot nonchalantly on the way they were headed.

There was a pair of coyotes that had their den in the horse pasture one spring, so I saw them most days, or at least early mornings when I went to get our horses. Jeff had a big roan mare at the time running with our horses, and often it was that mare that I caught to ride back bareback, running the saddle horses home. I'd go by the coyote den to see the four pups out playing on the mound in front. They were as curious to see me as I was to see them. Mom Coyote didn't like our glances back and forth and would order her pups down the den. They obeyed. When they were old enough to run farther from the den on their own, I didn't see them anymore. However, their parents gave me one of my worst scares while herding sheep.

It was spring lambing time, and I was herding several small groups of ewes and their lambs at a time on horseback. The sheepherders on foot were taking only one bunch. I would gallop back and forth between my three groups to keep them moving in the right direction. I'd just finished checking the other two flocks. The third flock was far enough apart, and grazing in a draw, to be out of sight from the others. When I loped back on this overcast, drizzly day, Mrs. Coyote had already made her kill. Evidently, she had watched me, out of sight behind a cactus, waiting for her chance to pounce upon her prey. A ghastly, tight feeling shot through my chest when I saw fresh red blood on the outstretched body of a lamb. Flying off Ginger, I ran to the lamb and manipulated its limp body with the hope that somehow it might not be dead. I was so appalled, as this was the worst thing I could have let happen. It is a rancher's code of ethics, instilled from parents and neighbors, that the animals come first. One must be ever alert to keep them safe, and I had not used sufficient caution. The folks depended on every lamb to make a living. That evening when I got the three little flocks home, separately one by one, I had to tell Mother and Dad about the loss. I knew they wouldn't punish me, but they would be disappointed. To have lost any bit of the trust they placed in me was my own worst punishment.

Jeffery and I Rope a Coyote

Mother's brother Will and his wife Ena visited the ranch a few times and brought gifts well-suited to Jeffery and me. When we were in early high school, Cathy, a cousin one year older than Jeffery and me, visited from Iowa. Cousin Nathan had visited the previous year, and we had initiated him into calf and horseback riding. Taking young visitors riding was our main source of entertainment for them, as well as riding calves in the corral and of course getting bucked off, frequently into some wet cow pie. We three cousins were riding in an area of the ranch called Cactus Flat when we came upon a young coyote, maybe a year old. Jeffery and I decided we'd show off to Cathy and rope the coyote and give her the pelt to take back to Iowa.

Neither of us had previously roped a coyote. It turned out to be much more difficult than we had imagined, for the coyote was not only very fast but extremely good at dodging behind and around the innumerable cactus in that area. Time and again we missed. Neither of us were any good at getting our lasso around that fleeing coyote's neck, as somehow it wasn't at all like roping a calf. Naturally, we didn't consider giving up, and in time we did get the coyote roped. But then what? We'd promised Cathy the pelt and we'd finally caught the coyote, but it was very much alive, leaping and snapping at the end of the rope.

I always carried a pocket knife, so getting the pelt off was not going to be a problem, but how were we going to kill it and not get badly bitten? In self defense, the coyote might at any moment turn on us and attack. Persons that want to show-off and don't first consider the consequences sure can get in a huge predicament. At last, and fortunately, the coyote crawled partly under a cactus bush to get away from us. We found some rocks nearby and hit the coyote on the head knocking the poor creature out. We put it behind my saddle to take it home to skin for Cathy to take its pelt back to Iowa.

In later years, government trappers put out scented cyanide capsules to lure coyotes by the scent, and once taken into their mouth, it would kill them. After a number of years of this eradication practice, the singing from the buttes at night ceased, the packs were seen no more, and those remaining became much more leery. Much of the charm of the prairie went away with the decline of the coyotes.

Prairie Dogs and Their Towns

When prairie dogs perceive danger approaching, they have the habit of sitting upright, very straight on the mound of soil in front of their den, front feet up at chest level, tail straight out behind, barking away with their little tails flipping up at each bark. They'd keep this up until the intruder got too close for their comfort, then flip, and down the hole they went. Those farther away from you would continue to bark excitedly. When I approached them very fast on a horse, they looked like a row of dominoes falling over, with the closest prairie dogs "falling" down the hole first, followed rapidly by the line of others. I never tired of watching them, and I often waited quietly to see how soon their curiosity would get the better of them and a head would pop above a mound to check if the coast was clear yet.

When I was older and ventured away from home for miles, I discovered one prairie dog town that particularly fascinated me. It was also the largest one around. It was located on the mesa-like area above Cactus Flat, above where an old log cabin was located. There were always a good number of prairie dogs sitting out on their mounds barking away, visiting, I suppose, with their neighbors and relatives. Seeing me ride up especially excited them and increased their chatter. I'd pull up Laddie and sit there observing the goings-on of the town for a long time. In those young years I'd not yet felt any rush for time. If one stayed still long enough, sometimes they'd scurry away from their mounds to search for things on the ground. As sparse as the grass was, I wondered how the colony found enough food to go around.

Sometimes I got off my horse and looked down the holes. What I'd heard about rattlesnakes living in the holes with the prairie dogs is true. Or let's put it this way, twice I saw a fat rattler in the entrance of a prairie dog hole, and when I attempted to kill them, each promptly slithered back down the hole. This was a fine situation for the snakes—a protected home and plenty of plump little baby prairie dogs to dine on. I felt sorry for the prairie dogs, but the snakes, in retrospect, were good to keep the overall prairie dog population down, so what grass did grow was enough to feed the colony.

Another live-in "guest" with the prairie dogs was the little burrowing owl. I felt this was a better combination, but actually it wasn't better than the rattlesnakes because they too lived off the baby prairie dogs; it only seemed nicer because the owls were so tiny and adorable. They let me come closer to them than a prairie dog would before they decided to duck down the hole, just in case I was planning to get a hold of them.

Other Wild Mammals of the Prairie

I liked to tease badgers. They were so much fun to bother, because they got extremely angry when pestered. The madder they became, the more I continued to taunt them, in a cautious way, of course. When herding sheep, sometimes there would be a badger out hunting. After chasing it around, the badger would try to dig a hole, very fast, to get away from me, but they never dug themselves out of sight. I always took it for granted that a badger would not turn on me and chase me instead. No badger ever did, but they would turn around to face me, snarl, hiss, and puff up, doing everything they could to appear as dangerous as possible to discourage anything from daring to touch or grab them.

One year Mother lost a dozen young chickens. Whatever it was got into the chicken house, killed them all, and buried several of them with just their feet sticking out of the ground. This puzzled Mother, who thought it must have been either a bobcat or a badger. She decided it probably was a badger.

We had skunks on the ranch as well, but they weren't an everyday occurrence like the rabbits. When trapping coyotes for their pelts, the folks found skunks in the traps a couple times, and I remember how stunk up they got releasing the skunks. They washed outside behind the house. Mother had worn a pink smock, and she buried it with their other clothes in the ground to rid the clothes of the smell. They didn't have many clothes, so couldn't afford to discard even the skunked ones.

Antelope were numerous, but there was a law that none were to be hunted or killed. One could be quite certain of seeing a bunch of eight or more antelope near the Jones reservoir, the Barry section, by the buttes, or just below the Woods pasture. It was always exciting to see a group jerk up their heads, flash their white rumps, flee off until they thought themselves safe, then stop to look back to make sure they were not being followed.

Little Varmints

Mice often got in the house. Mother kept a number of traps set and was good at knowing how to catch them—often just a little flour sprinkled on the trigger worked. If we saw a mouse in the pantry, a cat was brought in and the door closed. Things were moved around until we found the mouse, which ran, and the cat pounced on it. If we didn't bring in a cat, we just caught it ourselves with a glove. Mother hated it when mice got in the closet, where they could chew up a piece of quilt or some valued papers to make a nice nest. Well, not if mother could help it. Mother didn't even mind having a gopher snake in the house to keep the mice caught. But she did this later on, for I don't remember having a snake in the house when I was small.

I never saw any rats in the house and only once at the barn, so Mother must have eradicated them before they could multiply. But back to mice. We had a number of cats, and they took care of thinning the mouse population around the barns, but there were a few years when we hauled cane with the seed tops for the cattle's winter feed, and under the cane pile, where we had stacked it behind the barn, there must have been over a hundred mice. I caught and killed forty adult mice with my gloves before they scattered out of sight. There were enough mice and their babies in that cane stack to feed all the cats on the ranch and then some.

Wild Birds

I loved to watch the flights of wild geese and ducks that passed over the ranch in the spring and the fall. I would first hear their unmistakable honking and then look up to see a huge 'V' of geese high in the sky heading our way, becoming larger as they passed overhead. I'd just stand there, face turned skyward, transfixed, watching their every movement until they had winged their way out of sight and their honking faded into the distance. The time of their appearance also served as a weather forecast of a late spring or an early winter. The flights of migrating ducks were interesting to watch too, but never as exciting as the geese flying in formation. Both waterfowls found the Jones reservoir a convenient stop-over place, whether heading south for the winter or north in early spring. The fall stopovers seemed to be less hurried to be on their way south, lingering a day or more longer. At migration times I'd ride to the Jones reservoir sometimes just to see if there were geese. Some would be around the edges searching through the pond weeds for feed, and others swimming. If I was careful not to get too close or come in fast, they wouldn't take flight and I could enjoy them. Hunters came out from town in the fall to hunt geese and ducks, so they were always leery of any humans.

Killdeers were fascinating. They'd act "hurt" and limp off just barely ahead of me, so I felt sure to catch one, but somehow I never did, for they managed to keep just out of my reach. I enjoyed watching a particular pair of killdeer that hung out around the place or down by the arroyo, which they particularly liked after it had flooded. Killdeer "nests" were just eggs on the ground, and most years I'd see the young, the cutest little things, looking just like miniatures of their parents. I so wanted to hold one, but I had absolutely no chance, as the killdeers are fast and tricky. I never tired of hearing their call "kill deer, kill dee, kill dee" over and over. They'd give their call rapidly if taking flight, but usually they ran on the ground.

A meadowlark singing at dawn.

Meadowlarks were probably my favorite of the ranch birds. I loved them all, but the meadowlarks always awoke early in the morning, almost before the dawn. At four o'clock, when I went into the East pasture to get the saddle horses, barely able to see my way and wondering if I might have missed the horses, the meadowlarks would already be singing, loud and clear, what sounded like "Oh say wait a bit" or maybe "readin', writing, and 'rithmetic". I wasn't alone, because the meadowlarks were already up and singing joyfully. Their call cheered my heart. They were strikingly beautiful birds with the bold black 'V' across their chests.

Horned larks also lived on the prairie nearby. Once in a while I saw their nests, which were just round cup-like indentations on the ground. They were ground feeders, eating the seeds of grass, but I often noticed them near ant hills, and I wondered if they ate ants. Their gray color and markings blended well with the land, and their distinctive "horns" made them an emblem of the prairie.

A pair of owls always lived in the canyon and roosted on two huge rocks that jutted out from the wall near the top of the cave. Their hunting territory included our ranch barnyard over a mile and a half away. If something happened to one pair, a new set immediately took over, since the canyon was never without its pair of owls. When riding or walking up the canyon, they'd see me coming and fly farther up the canyon and out of sight, likely landing in a juniper tree. Seeing the owls fly soundlessly up the canyon was a spectacle I enjoyed and an interesting feature to show to visitors. The jutting rocks they perched on were coated white from their droppings, and the cave floor was scattered with the bones of small animals. They had a nest for their young on the cliff of the cave, and some years we saw the owlets.

One bad things about the owls was that they liked to eat young turkeys. Mother raised turkeys as part of our income, and they were kept in a small fenced pasture we called the "turkey yard". In this area, she'd planted a hundred trees given out by the government for windbreakers. It was in these trees that the young turkeys liked to roost, and for the most part they were safe, as no coyotes came into this enclosure. But the owls slipped in at night, especially on moonlit nights. They landed on a branch with the turkeys and sidled up to the one they intended for supper. They would eat the breast meat and leave the rest of the body, which greatly upset Mother, who built a shed to enclose the turkeys at night. The owls even got two kittens my parents were given to raise for mousers when they first came back to the ranch. One night we heard a scream from the kittens and found the owls had taken both of them from by the big clothesline post.

English sparrows were not one of our native birds, and for many years we did not have any, until at last they moved in. They could carry disease and also ate the grain and feed of our livestock, so they needed to be eradicated if possible. They roosted at night in the rafters of the barn and were catchable if one approached quietly. Many nights I made English sparrow raids.

We often saw mourning doves behind the corrals and in the arroyo, giving their plaintive calls of "coo coo cuk coo" over and over as they walked around while slightly bobbing their heads. They were rather tame and didn't attempt to take off when I approached unless they thought I had intentions of catching them. They were actually an outer barnyard bird that didn't come into the corrals often, though sometimes they came between the house and the barn. They liked places where gourds grew, such as along draws and arroyos.

Crows didn't come around the barnyard much, but they were mostly seen around the canyon or where something had died. They were talkative. I listened to their "caw, caw" and various other calls, curious about what they were telling each other. One crow always remained on guard for protection against any predators. They built stick nests near the top of big cedar trees.

Turkey vultures, or buzzards, could be distinguished quite easily from the crows even though both were black. The buzzards were bigger and didn't have feathered heads, but the top part of the neck was bare and reddish. They were quite plentiful; you'd see them circling in the sky, wings spread, probably trying to pick an air current with the smell of something dead. If a cow or other animal died on the range, by the next day there would be several buzzards on it, tearing it apart and gorging. Usually around six to eight buzzards were present, seldom more, rarely as few as a pair. They were so intent on what they were doing to a dead cow that they hated in the worst way to have to fly off just because I came around, but they considered it not safe to stay either.

Thunder hawks were an exciting evening attraction. During the day, they mainly stayed on limestone ridges, where they laid two eggs on the rocky ground without any nest. But toward evening, their favorite time of day, they'd swoop down close over me and give out a loud boom, or thunder, as their name implies, made by air in their wing feathers while diving. They'd thunder other than just around me, but having something to boom over apparently was more exciting for the hawks. Once, I stole away with two of their babies. They usually had only two at a time. I named them Thunder and Lightning. As is so often the tragedy of captive wild animals, seldom do they survive in captivity. First Thunder died, and a week later Lightning died also. I wished then that I had left them up on the ridge with their parents.

Other birds seen on the ranch included black headed juncos, which appeared in huge flocks, often in winter, never staying for long. Quail lived on the prairie in large coveys, usually several parents with all their young. Men from Fowler, after roads were built, used to come out to hunt the quail. This made Mother very unhappy since quail, although also vegetation eaters, consumed grasshoppers. Roadrunners were scarce, or at least seldom seen, usually only in Cactus Flat. We had all three kinds of blackbirds, but the yellow-headed and the red-headed were only passers-through. The black-headed ones, with their distinctive white eyes, stayed around the corrals. I made a pet of one of them, whom I named Prince. He remained friendly after turned loose in the barnyard.

Swallows lived in the canyon. Their colonies were large with jug-shaped, mud-daubed nests attached side by side to the overhanging rock ledges of the canyon, always busy darting in and out of the small round entrance holes of their nests.

Listen to the Mockingbird

By far, the most beautiful singers on the ranch were the mockingbirds. They had a large repertoire of all the other birds' songs, including those of the barnyard poultry, even cheeping chicks (so real they fooled me several times into looking for the "lost chick"). On top of that, each individual mockingbird had its own special song. At nesting time, the male sang nearly all night long. Some summer nights were mildly warm with only a gentle breeze and moonlight bathing the countryside in its soft yellow glow. How loud and clear the male mockingbird sang on such glorious nights; it was worth lying awake just to hear him sing.

When tending sheep around the canyon, often near cedars, the mockingbirds would be singing. I'd thought about getting baby birds, but didn't, as Mother had warned me that the other babies might be taken by a varmint if I were to touch the eggs or the babies. But one year temptation overcame me, and I brought two young mockingbirds home from a nest in the cedars. They were still quite small and not frightened of us, and considered us parent birds. Mostly Mother cuddled and fed them. I tried to help by bringing in lots of grasshoppers every day and other proper food for baby mockingbirds. In Walsenburg, there was a store that sold cans of mockingbird food, which we bought. It was a crumbled grasshopper type of food.

While small, they lived in Mother's apron pocket and, like baby kangaroos, they went everywhere Mother did. They were a male and female; Larry and Hallie we named them. As they got older, they'd sit on our shoulders, but their favorite perch was on top of our heads, where they'd spend literally an hour searching around in our hair. Actually, it felt nice and I enjoyed it. We never caged them, and they had the run, "fly," of the house. Their dropping were fairly solid, so were not messy. Hallie was sweet and sang some, but Larry became a great singer and had quite a repertoire, some of his songs being unique to a "people bird" that the wild mockingbirds didn't know. They were both beautiful birds and loved our family, but tragedy has a way of coming, no matter what.

Larry was the first to go. There was a stray cat that Mother didn't even like or want around. She attempted to get rid of it, fearing just what happened might. On a trip to town, she put the stray tom cat in a gunny sack, tied it shut, and put the sack on the floor board below the front seat. The grain mill in Fowler was good about taking in cats, as they needed them to keep the rats and mice out of their grain mill. After a thirty-five mile ride, tied in a gunny sack on the floor board of the car with no way of seeing where he was going, the cat was left at the mill. Some four days later, to mother's utter amazement, the cat was back at the ranch. And not long after that, he jumped through one of the window screens at the birds, the screen broke, and the cat got in and caught Larry. Hallie lived on for some time, but she loved to play with rubber bands as if they were worms. She swallowed a rubber band, and it didn't digest in her craw, filling it too full for other food. After she died, Mother discovered the rubber band.

Capturing a Red-Tailed Hawk

Red-tailed hawks gliding on wind currents high overhead while searching the ground for prey were a nearly daily sight. One time I was riding in the northwest part of the ranch near an old landmark cedar we had named Letitia. A pair of red-tailed hawks were using a stick nest, built the year before by crows, in the top of this tree, and visible just above the rim of this nest I saw three cottony-white, woolly-clad chicks with fierce eyes and sharp beaks. They were a couple days apart in age, and I wanted one. The biggest was the size of a grapefruit. The parents were gone just then, so I climbed up the trunk and through the branches to take one out of the nest. The captured chick was vicious and struggled to get free with its beak and claws combined while shrilly screaming for its parents. Its folks must have heard its screams, for in an instant, before I could get down from the tree with the fighting baby, they were swooping down on me with savage, piercing war cries. I was scared to death of what they'd do to me, and I jumped on Laddie, who took off at full gallop before I had barely more than one foot in a stirrup. I managed to hold on to the young hawk, who continued to scream with both parents following, swooping down close, too close! They looked mean and fierce, and I knew they meant real business if they could get a claw into my head. Laddie, afraid for his head too, flew for home like the devil was after him, never slowing down until we were in the barn with the door shut. The hawks, seeing no chance of retrieving their baby, had circled back toward their nest and the remaining young.

I took my prize to show Mother. I can't say that Mother was too thrilled with the new addition to our family, but she did advise me on the young hawk's care and diet. I began hunting mice for him. He lived in the house with us, accepting us for parents after several days of snapping his beak. He loved to sleep and snuggle with me on his towel under the covers at night.

We had George until he nearly grew up. He was extremely intelligent and was named after George Washington. He knew his name and came when called. As he got bigger, he spent the days outside in the garden. I tried to teach him to fly and hunt to no avail. I liked George, but Mother was leery of him. She said he might catch her hens when he got older. When he got older, he started spending time in the barnyard, and he'd perch on the woodpile near the house and eye the chickens as they scratched around. George was somewhat of a playmate; he'd go with me around the corrals and down to the arroyo; we needed each other's company I guess. But the day came when he flew down from his perch on the woodpile and seized a hen. Mother said he had to go; she'd warned me. George was my responsibility, and Mother offered no help. Sadly, I had to kill George, as he wasn't capable to fend for himself on the prairie, as he depended on us too much for food. With a hatchet in hand and George under my arm, I sadly walked down to the arroyo telling him I loved him and how sorry I was. With one sharp, quick blow I took care of my responsibility, which was followed by a lengthy cry. Afterward the place seemed empty without George around.

Cactus

The cactus is a symbol of the prairie and certainly one of its dominant plants. All cactus have lovely flowers, but my favorite was the velvety magenta to red flower of the tree cactus with its yellow center. Prickly pear cactus sprawl over the ground and have yellow or greenish yellow flowers. The prickly pear with its very long thorns was more dangerous, since it was hard to walk without stepping on them. Their pears were great for making jelly, once you got past the thorns and fuzz.

The little pincushion cactus grew solitarily on the prairie among the grama grass sod. With its little bright pink, pointed petals, looking like sea anemones, it vied with the tree cactus as my favorite. Seeing one in bloom tempted me to dig it up to bring home for our cactus garden located between the cellar and the clothesline. Mother had arranged some rocks there making it a rock cactus garden. Another little pincushion cactus grew near the canyon that had almost green flowers. A larger variety of pincushion cactus, shaped more like small barrels, grew in clusters in the canyon. It had different colors of flowers: greenish yellow, pink, or red. We had two of these in our cactus garden as well.

Yucca, often called soap weed or soap plant, has fibrous roots about an inch thick. Mother would get a small piece of its root, pound it with a hammer, and use it for soap to shampoo our hair. With soft snow water and soap weed, our hair just squeaked it was so clean, and very soft afterward. This was the best shampoo of my life. The flowers of the yucca are a lovely creamy white, often referred to as candles of the Madonna or candles of the Lord. Red racers startled us a few times by camouflaging themselves to resemble the yucca's pink bud on its stalk. Oftentimes a jackrabbit would make its little indent form in the ground under a yucca. A few times I discovered a stink bug impaled on the spear of a yucca's pointed spine. I loved the yucca as a prairie friend. There were special places where it grew best and formed colonies.

When the temperature was below zero and I was very young with Mother on the prairie herding our flock of sheep, she would search for old dried yucca roots, when no cactus wood was around, to make us a tiny fire against an arroyo bank to warm our hands and feet.

Cedar Trees Can Be Best Friends

If asked what my favorite tree was, the answer without hesitation would be the cedar. It is impossible to imagine what outdoor life on our ranch would have been like without the cedars that grew on the shale limestone ridges and rocky areas. Cedar trees were a part of me, one of my childhood "companions". Children who have brothers and sisters, or neighbor children to play with, or the modern distractions of television and ample books, might never think much about the various kinds of trees, but when you are by yourself on the vastness of the prairie, often for hours at a time, they become important. You are not "alone" because these trees are nearby and can be counted on to be there for you.

When visitors came, and we wanted to show them around and have a good time, invariably we chose the canyon for its spectacular rock formations and the cedars, which provided a pleasant contrast to the flat prairie grassland. Usually we took a picnic lunch to eat in the shade of a tall old cedar. Iowa relatives and other friends enjoyed hunting for arrowheads, which were common in cedar-clad areas. In the cedars above the canyon several scattered stone circles remained where tepees had once stood.

In summer, the sun bore down intensely on most days. When young herding sheep, if anywhere near the cedars in the heat of the day, the sheep would all crowd tightly together under cedars for shade. I would lie stretched out to cool off under one of the taller, gnarled old cedars and look up into the deep blue sky, often watching a red-tailed hawk gliding on the upper wind currents. In wintertime, when herding sheep on foot or checking on the cattle by horseback in freezing weather, I'd stop to cuddle against one of the younger bushy cedars, close to its trunk, as a break from the wind. Often icy winds drove sleet right into me and chilled me to the bone. Stomping my feet for circulation on the lee side of a bushy cedar may have saved me from hypothermia at times. The cattle were wise and stayed around the cedar breaks, too, during storms. They'd stand hunched up, head down, as close as possible into a cedar to stay warm.

Dry cedar limbs were abundant on the ground, as in previous homestead years, to prove up their land, people needed lots of posts for fences and corrals. Many wagon loads of cedar posts had been cut and limbed. The miles and miles of homestead fences owed the cedars for their existence, some of the mighty old cedars being used for the big corner and gate posts. Another use was for pickets. We had a picket fence corral when I was young. Other ranches, like the Red Top and Sanford ranches, had large round corrals made of big, very tall pickets, at more than eight feet tall, high enough to keep wild horses from attempting to jump over them.

With Paint, showing our picket corral fence.

We named some landmark cedars after people. A small cedar tree on the knoll back of the barnyard was called Barbara. The big cedar where I got my pet red-tail hawk was named Letitia. The tree I liked to rest under when riding by the canyon was Bill Black, after my dad. A tree at the upper end of the Barry section was Jerry Simpson. A tree on the road to the cedars was named Bill Henderson.

In certain places pinyons intermixed with the cedars, but they didn't grow everywhere the cedars did, only in certain areas. They had pinyon nuts, which would have been a treat, except that animals were always ahead of us. The pinyons had a resin that oozed out from a tree wound and hardened. This was about my only chewing gum, and a great treat, except a few times it was a disaster. For some reason, it would get stuck all over my teeth and mouth if not started correctly, and it was then very hard to get rid of. A branch from a pinyon was what we used once a year as our Christmas "tree".

A pinyon was occasionally barked to the extent that it killed the tree. That was the work of porcupines, who liked to dine on pinyon bark. There was a law that people were not to kill the porcupines to save the trees, for it was said that if a person was lost, porcupines were one of the slow animals that a person could kill for food to keep from starving. I wondered about the logic of that law as a child, reasoning how it would be difficult to find a porcupine in the first place, and if so, how would I kill it and not get hurt by the quills. If I succeeded, how would I manage to get it skinned?

Nowadays when visiting the ranch, I take time to go see some of these special old trees, as if they are old-timers—friends. They were old over eighty years ago when I sought their shelter; they're old now, as if somehow they will be there forever. These dear blessed cedars, you have meant so much to me.

Visiting a special "old-timer" cedar.

There was a time after the dust storms that the government gave ranchers trees to plant for wind breaks, hoping to break the force of the wind that ripped over the prairie. Mother accepted one hundred trees of several kinds, which she planted in a fenced hollow area northeast of the house. She chose the hollow where moisture would remain in the ground longer. The trees were cedars, elms, honey locust, Russian olive, plum, yellow pine, and choke cherry. Most of the trees lived. Some grew for a few years, many around thirty years, and others are still living eighty years later (such as the cedars, now thick and bushy). Over the years Mother turned the area around the ranch house into an oasis of trees, bushes, and vegetable garden, so that her family and others might feast their eyes on greenery in a sea of dry prairie.

Flowers and Weeds

When living in a small town most people are known by name and spoken to as they are met on the street. When I was young there were no people around other than my parents and a rare visitor. So being out on the land, it was the flowers and plants and also the animal life that were noticed as the individuals they were. They kept life interesting.

Only a few plants were for eating: dandelions, lamb's quarter, pig weed, and Russian thistles, better known as tumbleweeds. Mother would pickle the early, young thistles. When they matured and became huge tumbling weeds in the wind, they made excellent gully stoppers to slow erosion. Mother steamed the edible plants to cook them lightly before we ate them, but they could also be eaten raw. Currants, gooseberries, and chokecherries grew in the canyon. We ate the currants off the bush, but the gooseberries were only good for pie and the chokecherries for jam. There were the bad or poisonous weeds, such as whorled milkweed, death camas, larkspur, prince's plume, a brown weed, gourds, and cockleburs, which got in our cows' tails, horses' manes and tails, and the sheep's wool.

Spring flowers were special. The first to arrive were the inch-high, white Easter lilies. Next came the low yellow violets. A few weeks later, many more flowers bloomed, including pink phlox, purple verbenas, and primroses. Blue lupines, pentstemmon, paintbrush, and four o'clocks grew on the limestone ridges. After Highway 10 was built, sunflowers, creamy colored mentzelia, and floppy white poppies grew in the bar pit. Bushy prairie plants called rubber plant and rabbit brush bloomed with yellow flowers. Low-growing sage was very prevalent. Ground vining burs, like puncture vine (we called tackburs) because each nut-like bur has five tacks. Their sharp points were a nuisance. Sandburs were worse, encased as they are with tiny burs. Beggars lice were covered with tiny round burs that matted one's socks.

Prairie Relics

In later years I never saw them anymore, but when young it was exciting to find buffalo skulls protruding from the arroyo banks. They were two or more feet down from the top ground level. It was not that many years since the buffalo had roamed over the prairie, so I wondered why they were buried so far below the present ground level. I also found a few curved buffalo horns on the ground, likely carried away and left there by animals, which I brought home for souvenirs.

Beyond the house about half a mile was a small black volcanic butte. I loved to climb around its sides and on its black volcanic boulders. Here I found white chunks of gypsum with a very smooth cleavage. I always brought some of the whitest and most beautiful chunks home.

A little farther to the north was a low, cow-pie shaped knoll with a flat top of blackish, slate-like soil. There was only one other little knoll at the Buttes like this one. Here we hunted for sharks teeth. In the early years, we found big, inch or more long molars, but in later years the big ones were hard to find, and we found mostly small molars and pointed teeth. Poor little knoll, it was continually having to give up its treasures to us and other visitors we brought for the excitement of hunting sharks teeth.

Haystack Butte at one time had large geodes eighteen to twenty-four inches in diameter, but these beauties were soon depleted as people from town took them to adorn their collections and yards. Little by little all the big and best ones disappeared, leaving mostly the similar six inch to eight inch geodes and broken ones with lovely, almost white, crystals inside. Haystack is also where I found the most iron concretions which had eroded out of the limestone, some of which I would collect and put in my pockets to take home. The little ones were nearly an inch long and the big ones over two inches long.

These various relics provided much happiness and hunting adventures during my younger years.

Arrowheads and Indian Artifacts

Several artifact collectors from Fowler wanted to come to the ranch to hunt for arrowheads, which Mother was kind enough to allow, but it is a shame that much of the ranch's Indian remains went to outsiders which should have stayed on the ranch. In addition to that, ninety percent of what we found ourselves ended up being given away. Visitors would come, and seeing the arrowheads Mother and I had found, they wished to have an arrowhead. Mother would give them one, usually the one they wanted. In the end, when most of the good ones were gone, she decided to glue the rest down under glass in an organized arrangement so people would not try to ask for one. That fraction of our overall collection is all that remains to our family.

In the canyon cave there are huge boulders whose tops have been chiseled and smoothed into large "bowls" for grinding meal. The Arapahoe, Comanche, Kiowa, and Ute all lived on the plains of southeastern Colorado at one time or another, living as nomadic hunters and gatherers. My mother thought that the Indian remains in our area belonged to the Arapahoe, but we never had this confirmed. The Indian ladies must have sat in the cave to do their work, where it was pleasant in the summer and offered shelter in the winter. Some of the smaller "bowls" have been taken, but all the bigger ones are left. It would be nice if the full assortment of artifacts were still there, just like it was when the Indians used them. We all have the urge to collect, to take, until what had been interesting and thrilling in nature becomes not nearly as exciting on the shelves of a collection.

Mother had many grinding bowls which she found on the prairie and lugged home, only to give them away to someone who asked to have one. She always found the bowls upside down, looking like an ordinary slab of sandstone. Possibly, their upside down placement was done by Indian ladies to disguise them from being taken or discovered. They weren't located in rocky areas but mainly out on the prairie sod, where a tepee might have been pitched. Mother found grinding stones as well, usually not near the bowls, and rarely on the prairie, but mostly above the canyon rim among the cedars where there were rings of rocks where the Indians had had their camps. Other bigger stones showed signs of having been their campfire rocks. With the passing of years and with the rocks being moved by visitors, even by our sheepherders, the evidence of the circles has gradually disappeared.

Mother found two more-modern steel arrowheads in the east end pasture. The Indians had started to use them after their contact with the white people that came to the area. Mother's favorite artifact, which she found in the barnyard, was a tiny, red carved owl. But it must have been lost by one of the men who boarded with us during the construction of Highway 10. It appeared to be Aztec in origin and not made by the Indians of our area. I still have some tiny, perfectly shaped bird arrowheads that Mother kept in a small round container and fortunately didn't give away.

Mother's advice to me to "always watch the ground where you step" (to see a rattlesnake before stepping on it) was also helpful advice for spying arrowheads. I once found a nicely shaped pink one right in a dusty cow trail. It's a wonder it wasn't chipped or broken with all the cattle hooves that must have trodden over it. Probably the thick dust was a sufficient cushion. Another time, while galloping on a horse on the east side of the canyon, I spotted an arrowhead on the ground and pulled the horse up fast. After picking it up, I discovered it to be a beautiful specimen.

Once on the south end of the Mobly ridge near some old cedars, I found six arrowheads and lots of chips. That was a thrilling windfall. Some were flint of different colors and a few obsidian. I pictured an older Indian man who didn't go out hunting any more contentedly chipping the arrowheads while sitting comfortably against the trunk of one of the cedars. I wondered why the arrowhead maker had left so many finished arrowheads in one place. The south end of the Mobly ridge was generally a good vicinity to find arrowheads. The other best place to find arrowheads was above the canyon rim on the west side among the oldest cedars. These trees had probably been fully grown at the time the Indians lived there. Bill Black Cedar was the tree under which the most were found. I always managed to find at least a chip, if not a broken arrowhead, and at times a whole arrowhead or two.

This was where we brought our visitors to hunt arrowheads, since they would have a good chance of finding something. On the prairie itself, now and then, you'd see a nice one on the ground, especially after a hard windstorm sent the dust sailing, uncovering arrowheads that had lain hidden just below the surface. I also found arrowheads on a few occasions near anthills, where the ant activity had probably uncovered them. I was always proud to come home with an arrowhead to show Mother and Dad and to add it to our collection.

We found as many skin scrapers, used for skinning game and preparing hides, as arrowheads. The scrapers were usually made of a dark or dull colored stone, two to four or more inches long, about the width of a knife, and notched for the thumb and finger to fit around while skinning. As they were not as beautifully chipped or made of pretty colors of flint and obsidian like the arrowheads, people did not often ask Mother to give them a skin scraper.

The crow bar Dad brought from Idaho. He purchased his Idaho ranch from his dad in about 1894 and the bar was already in use. It helped to dig many a post hole and to pry out the underground cedar stumps when we gathered stove wood, among many other uses. Mother's old .22 rifle was seldom used, but in the early years Dad shot rabbits for our meals and the coyotes they trapped (not to mention skunks that also got in the traps). Dad always kept it cleaned and oiled, ready for use if needed.

Part VI

More Memories of Mother and Dad

Mother used the coffee grinder all her homesteading years, not just for grinding the morning coffee but for grinding many grains and seeds. I particularly remember our cooked morning cereal she made by grinding wheat purchased cheaply for livestock. She said it had more vitamins than the Kellogg dry cereals.

She had three flat irons of different weights. When not in use they were set on end at the way back corner of the cook stove.

Vignettes of My Dad

Dad was always proud of me, but he was not an affectionate person any more than my mother. I don't recall him hugging me or putting an arm around my shoulder, but I knew he liked me. Mother said that when I was born, he exclaimed, "Damn, what a pretty baby!" All his grandchildren, except for one, were older than me, though I don't think I felt like a grandchild to him. I definitely was very different from his first daughters when they were little girls. They were "done for," "spent on," and "spoiled" with lots of money, he lamented. Dad was complimentary of me, saying that I was never any trouble to him. My childhood was certainly far different and the opposite of his first family. I grew up without money in the Depression and without playmates, so I learned to take pleasure in working, which I considered more fun than playing. I was Dad's helper with anything that he was doing outside.

In the house, one room served as our kitchen and our living room, and the other room was our bedroom. Dad's place in the kitchen was in the corner by the wood box near the stove. His chair was there, just a plain four-legged chair, but Dad always tilted it back on its back legs against the wood box. Mother didn't approve, saying he would break the chair, but it never broke. But he had the annoying habit of chewing tobacco, and therefore he spit often. Being near the stove, he intended to spit into it while holding up a lid, but seldom did it all go straight in. What hit the edges of the hot stove naturally scorched, and to Mother and me, it smelled horrible.

My spot in the kitchen was standing up with my hands behind me, holding onto the smooth, round chrome bar in front of the cook stove. It was a nice warm place to be, and the bar was never hot enough to burn my hands. Mother's place, if not working, was sitting in a chair at the table reading. If I had lessons to do, I sat at the table too, by the light of our kerosene lamp.

We didn't have a lawn. None of the neighbors had lawns. Water was too scant and precious, and time that would have been spent on keeping a lawn up was much better used in other ways. But we did have an alfalfa patch about 60 x 75 feet in the yard. Mother used a butcher knife to cut the fresh, green alfalfa as green feed for the chickens, a treat for sick animals, and sometimes to make tea for us.

Once Mother was away for the day. I wanted to surprise her when she came home by having the alfalfa neatly mowed. I got down on my hands and knees like Mother did, held a clump of alfalfa in my left hand, and with the butcher knife in my right hand cut the alfalfa about two inches above the ground. I was mowing right along, probably having cut at least half the patch, when I carelessly sliced a deep cut on my left thumb. It bled in huge drops, making it appear very serious. I hurried to the house for Dad to fix it. What I remember most was how scared Dad was, and how desperately he wished Mother was home. She was the nurse; she'd know how to fix it. In spite of feeling helpless, Dad did manage to get the bleeding under control and rigged a bandage for my thumb.

Dad had only completed third grade, yet although he depended on Mother to do the reading and writing for the family, he could read and write quite well. Other than his signature, I only saw his writing

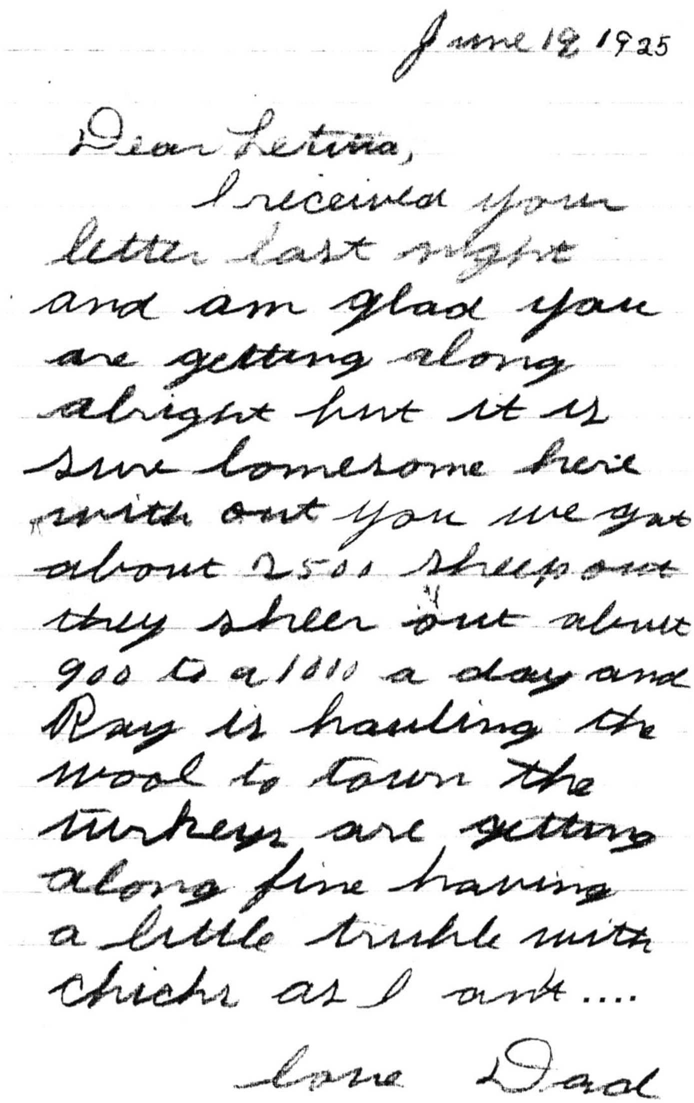

June 12 1925

Dear Letitia,

I received your letter last night and am glad you are getting along alright but it is sure lonesome here with out you we got about 2500 sheep and they sheer out about 900 to a 1000 a day and Ray is hauling the wool to town the turkeys are getting along fine having a little trouble with chicks as I aint....

Love Dad

in letters Mother saved that he had written to her when she went to Iowa for her father's funeral. Each letter mentioned how lonesome he was, that he'd sure be glad to see her back, that he'd lost five pounds, and that even the dog cried for her every day. In these two-page letters his grammar and spelling were nearly correct with nice, very readable handwriting. Dad never had a pair of glasses.

We had a gray-colored, second-grade reader, which had many good stories to enjoy. Dad would read these stories to me, and then have me read them back to him, helping me with the words I didn't know. We spent considerable time together with this book as I learned second grade reading with my dad. He would read to me a few of my favorite stories that I wanted to hear over and over.

When my baby teeth started to come out, it was Dad who was the official tooth puller. Sometimes he'd use his thumb and index finger, but more often he pulled a string tied to my tooth, or sometimes he tied it to a doorknob. Mother was the tooth fairy who came in the night. For some time, I believed in the fairy and the little things she left me that I'd not previously seen.

Some years later when we had a deck of playing cards, Dad taught me how to play casino. It must have been one of his favorite games, or at least one he thought easy enough for me. How excited I was if I got the deuce of spades (little casino), and I was really thrilled to capture big casino (the ten of diamonds). It was a pleasure playing casino with Dad.

One year Dad had a family reunion; probably it was 1936 for his 70th birthday. All four children of his first family, my half brothers and sisters, came to visit. Charlie and Margaret and Mary came from California, and Violet from Reno, all traveling together. Jeff lived locally in Fowler. They had a marvelous time reliving the past as they visited for several days. It was nice for me too, to get acquainted, as I had been too young to remember them before we left California. Mother was amused that neither of dad's daughters could squat down low enough to use the coffee can as the commode at night. Charlie and Mary each visited Dad another time later.

Dad managed to fool Mother and me on April Fools' Day twice with the same trick. He went outside east of the house early in the morning, and as he came back, he called to Mother and me to come out quickly and see. "I just saw a strange, big bird back there; come quick and see what it is!" We ran out quickly, all excited, and then were disappointed as he laughed and said "April Fools'". He pulled the same April Fools' trick on us two years in a row.

I frequently teased Mother. Knowing that I had a playful nature, she generally did not scold me. Slipping up behind her and untying her apron strings was my favorite way to tease her. One day she had told me to quit it more than once, and I had not minded. Usually, I minded very well, but this time I continued playing. Knowing it was very bad not to mind, let alone not to quit, Dad grabbed a stick out of the wood box to spank me. I ran, and being able to outrun Dad, I ran as far as the arroyo before he gave up chasing me. Fortunately, he had put the stick away and calmed down by the time I came back.

Dad was a superb storyteller. In his younger years in the late 1800s in Idaho, the state was still very wild. Not only were the events of his early life exciting, but his vernacular and knack for telling a story made me feel I was right there. If Mother or I tried to repeat his stories, they turned out dull, as if lacking seasoning. It is a pity that his stories of some of Idaho's early pioneers were never written down for posterity. Dictation was necessary to take down his stories word for word as Dad spoke them. I took shorthand in high school, but when I prepared to take down his stories, Dad was no longer able to tell them as in earlier years when we listened spellbound.

Storytelling

On evenings when we had company or when my cousin Jeffery was visiting, we'd sit on the bench enthralled as Dad related playing with the Indian boys he grew up with. One of their games was butting heads, likely in imitation of the bull elk. He told another story about a time on horseback when he'd been followed by a pack of a dozen wolves and was lucky to escape alive. Another story was about his mother and sisters being down at the river washing clothes when they were warned of an Indian attack (which turned out to be false). Afraid they might be captured, they hurriedly threw their clothes and tubs on their wagon and fled to Fort Boise.

Dad had one special story we asked him to tell time and again. It was his ghost story, and a mystery that plagued and puzzled him all his life. There was a plot with a few graves where some emigrants had died and been buried. Several times as he rode by this spot, he saw two little old women walking around and around one of the graves. They were dressed all in black for mourning with only a white collar or breast piece. They would just keep circling this particular grave. Dad never heard them talking; they never seemed to even notice him, but just continued walking with their eyes toward the ground. They looked exactly like two, not very tall, old women. Dad didn't have the courage to dismount and walk over to them; besides he was barely able to control his horse, which always spooked and reared when it saw them. Sometimes they would be there, and at other times they weren't.

The following story was recounted by William Black to Letitia Black and written down by her.

Our Guest

When I was about eighteen or twenty, we used to take our cattle to Snake River in the winter time to winter on the grassy flats. There was less snow there than at the home ranch 75 miles away at Hood River. Snake River cuts deep into the lava beds in places making deep canyons, and in other places it runs close to the surface. In most places the rim rock runs about three miles back from the river bottom. From the rim rock on back are mesas.

Late one fall, about the first of December, Hunter and I started out with about three hundred head of cattle, our carry, and our chuck wagon. The first night out we camped about eighteen miles from home, and it began snowing. It kept it up all night. The next morning we had about fifteen inches of snow, and it was still snowing. We had to leave our camp outfit and trail the cattle onward, eating at houses along the way. The cattle were strung right down the trail. They were used to wintering there [Snake River], so we couldn't have held them back if we'd wanted to.

When we had the cattle located [at Snake River], Hunter and I took some pack horses back to where we'd left our wagon to get our bed rolls and what supplies we could carry. We started back to the herd, camping at nightfall in a deserted

> cabin consisting of one room with the windows and door out. We unpacked our horses, tied them up, fed them grain, and carried our bed rolls and grub inside. I built a fire in front of the door while Hunter mixed the sourdough bread, sliced the bacon, and put on the coffee. After finishing our supper, we put the bacon that was left in the Dutch oven with the bread, and set it in the cabin. We lit our candle, which was stuck in a bottle, rolled down our beds, and turned in.
>
> We'd just gotten to sleep when someone hollered. Hunter jumped up, went to the door, and answered. A man dressed in a long brown denim overcoat, his feet tied up in gunnysacks that were worn through at the bottom, came in. He said he was a sheepherder, and had been lost for three days without a bite to eat. We lit our candle, offering to cook him a warm meal. He said there was plenty left in the Dutch oven. He cleaned it up. Hunter rolled him out a bed, and we talked a while before going to sleep.
>
> The next morning Hunter jumped up early to make our breakfast. He exclaimed, "That darned sheep herder! There isn't a sheepherder here! He's gone! He must be crazy!" So I jumped up. The food was still in the Dutch oven, the bed was rolled down, although no one had slept in it. The candle had burned about a third shorter. We circled all around the house. The snow was deep and there was no sign of a track except those made by our horses coming in. He was a nice polite kind of fellow. He didn't want to put us to any trouble.

Mother always cut Dad's hair and trimmed ours when it got too long. I wasn't very old when I got the chore of cutting Dad's hair with a comb and our one pair of scissors. I was worried I'd make his hair uneven at first, but afterward I enjoyed giving Dad his haircuts. When shaving became difficult for him, and I was home, Dad enjoyed my shaving him.

The cook stove consumed a great deal of wood for cooking year round and for providing heat during the cold months. As fall neared, Dad, Mother, and I would go in the pickup to the cedars to gather and haul many loads of wood until the woodpile became stacked high for winter. We drove our full loads back slowly so no pieces jolted off. Mother always left a pot of lima bean soup on the back of the stove, knowing we'd come back starved. It was so delicious one bowlful

Getting a load of wood at the cedars.

was never enough. Going to the cedars for wood never ceased to be an exciting adventure. Once at the cedars the folks got busy gathering wood, but I always dashed around through the cedars first, exploring, before getting busy helping. While Mother was gathering branches and small wood off the ground, Dad and I pried out stumps. In earlier years, most of the big cedars had been cut and limbed for posts. It was their old dried limbs that Mother gathered and their stumps that Dad and I worked out of the ground. Once in a while one came out easily, but generally each stump was its own puzzle to remove because of the web of its deep and strong underground roots. Dad and I worked together with a bar and ax, using rocks or other stump pieces as leverage for the crowbar. We were very proud of our stumps. At home they were chopped for the stove, making excellent heat for cooking and a pleasant aroma. I still have that crowbar, which Dad had kept from his early Idaho days. It's one of my prized possessions, as it reminds me of Dad and our comradeship working together.

When the folks went to town and were returning, Mother gave a prayer of thanks about every mile we safely got closer to home. This time Dad was driving alone. The night was pitch dark and he was only five miles from home when the old Chevy quit. Eager to get home, he set off walking. Three times, after considerable hiking, he recognized a spot he'd been before! He was only walking in big circles. Mother worried that night as Dad didn't show up. He ended up sleeping in the truck and walking home at first light.

During World War II there was food rationing. When we got groceries we only bought what we were allotted. After one trip to town, Mother and I were in the pantry a few days later and were shocked to discover three cans of coffee on the shelf. The only way they could have gotten there was that Dad had purchased them without ration stamps unbeknownst to us and slipped them into the pantry. Although the folks only drank coffee for breakfast, he must have felt it a necessity this far from town to not run out. This was a great jolt to me, since it was the first time I'd ever known anything dishonest—that of exceeding our ration stamps.

One time the weekend slipped away. It was Sunday afternoon, time to go back to Fowler for my next week of high school. With November's short days, Dad was anxious to get off so he could be home before dark. "Aren't you ready yet?" he said several times. I was doing my own cooking and it was taking too long to get my week's food together and the wood chopped I needed for cooking it. We didn't get the early start Dad wanted.

Early the next morning Jeff came to tell me that Dad had had an accident on the way home. There is a deep gully with a narrow bridge about five miles from home. With his poor vision and dim headlights, he'd hit the railing of the bridge, which overturned the pickup and rolled it completely over into the bottom of the deep gully. Timbers from the railing were sticking through the radiator and into the floor of the cab. Soon a neighboring cowhand who lived west of us came along and saw the headlights shining in the gully. He climbed down and got Dad out of the pickup and took him home. Fortunately, Dad was not much injured. I felt bad about Dad's accident, because if I'd just gotten ready sooner, he'd have made it home before dark like he had wanted. After that Mother didn't let Dad drive anymore. She had to hide the keys, which made him quite angry. From then on I did most of the driving.

When Highway 10 was nearly completed, the government still had equipment on the site, and they hired my dad to be a watchman up by the Hogback. The folks felt fortunate to have some money coming in. It wasn't difficult work, but still it was lonesome to be without companionship. He wasn't much at reading, didn't write, and had no radio, so I can imagine the days and nights were long for him. The workmen were not around most of the time, probably because it was winter time.

Up until he finished, Mother and I went once a week to visit Dad. She walked and I rode Ginger along Highway 10. Mother always carried Dad some food. It was a long walk somewhat uphill and a good five miles. The two things I remember most are the sore protruding far out on Dad's lip and the time a speeding car swerved out and almost ran me over on the horse, which made the horse jump down into the bar pit where Mother was walking and nearly knocked her down. Mother was as angry as I had ever seen her. She was furious at the young driver for swerving at me on the horse, nearly hitting it. I was trying to apologize for letting the horse almost knock her down. But she knew perfectly well I couldn't help it and that the horse had instinctively jumped out of the way. The red car, driving much too fast for that rough road, sped right on leaving a trail of dust blowing behind. We were less than half way up to see Dad. Mother was so enraged, she fumed all the way up at that young driver, "Why he could have killed you!"

Dad never smoked or drank, but he did have the bad habit of chewing tobacco. He kept a plug of Horseshoe tobacco in his shirt pocket along with some toothpicks. In the end, the tobacco juice that ran down the corner of his mouth caused a cancer that grew as big as a banty chicken egg before it was removed. He didn't have very many teeth left by this time in his life, but those he had were his own, which he cleaned with a toothpick, frequently left hanging out the side of his mouth. Through the gaps in his teeth the tobacco juice would run out and down his chin, which was shaped so that he had a little crease in the middle where the tobacco juice got caught and dried. Due to long years of chewing, he eventually got a sore on his lip that didn't go away. It was during the time that he was a watchman that Mother and I noticed that the lump on his lip was growing bigger and faster. After he was finished with the watchman job, he had an appointment with a doctor our local doctor had recommended in Pueblo, where the cancer was removed. When he was ready to come home, I remember how much he longed to have "a good cup of green tea".

If only I had understood Dad with my grown-up eyes and comprehension. I made him weep once, and I've felt ashamed for doing it ever since. I was in high school at the time. He tried to tell me that he had put a lot into making the ranch. By this time, he was an old man in his eighties, and as such sat around the place much of the time, unable to put in much work. That's what I saw then, rather than the man he had been. So I replied that I could see with my own eyes all the work that Mother had done, and acted as if I didn't believe him, giving Mother most of the credit. Being twenty-three years younger than him, of course she had greater energy, yet I was comparing them as equals regardless of age. Poor old man; he was just trying to get a bit of understanding and a little friendship from his daughter, and I acted like I doubted him, seeing the value and work only in my mother.

Dad's Passing

During the last days of Dad's life, Blanche Bennett came often to assist Mother while she cared for him. Jeff also came part of those days. After Dad's spirit had departed, Mother bathed him and prepared his body better than any mortician would have done. She even knew how to set his eyes. The date of his passing was August 3, 1951.

Dad had a nice funeral held in one of the churches. Mother chose a sheaf of golden wheat with red roses to place on the casket, and friends had their floral sprays. Vestal Browning, from the grocery store where we always shopped, sang. I had not heard her sing before, and her voice sounded absolutely beautiful as she sang hymns for Dad. The neighbors and friends were there. One of the nicest things about a small town and community is the stalwart neighbors that stick by their friends, especially when there is a need. Afterward the hearse took Dad to the cemetery, and police closed Highway 50 with all traffic stopping in respect for the deceased. I remember crying over that because Dad would have liked knowing that even the traffic of Highway 50, which passes through Fowler, came to a stop that early afternoon for his procession as it proceeded, all the cars with headlights on, slowly to the cemetery.

Mother had chosen a spot in the cemetery close to some of Dad's old friends that had gone there ahead of him. It was a beautiful clear day. Family and friends bowed their heads as the casket was lowered. Mother had done her best. She was very tired, not just from that day, but from her long care of Dad, as he was not a light man to move in bed and his bedding required frequent laundering. Now we were heading back to the ranch and a chance for Mother to rest and try to keep her own health. It would be quiet with Dad gone and Vinson and I returning to California after another day's visit. I was sad Dad was gone, but he had lived a full, long life into his mid-eighties, and his early years had been productive and adventurous in the young state of Idaho in the 1800s.

Recently I came across the August 10, 1951, Fowler Tribune obituary of Dad. I'm entering it here since it reveals Dad's younger years before I was born.

Funeral Services Held Monday for William M. Black

85-year-old Pioneer Died Friday at His Ranch Southwest of Fowler

Funeral services for the late William Marion Black, husband of Mrs. Letitia Black and father of Jeffery Black and Barbara Black of this community, were held Monday afternoon at 2:00 o'clock. Mrs. Joe Fox, Mrs. Donald Buck, and Mrs. Murrell Shearer sang three numbers, "Home on the Range", "Nearer My God to Thee", and "In the Garden". Casket bearers were Wm. Henderson, Floyd Krieder, H. H. Knowlton, Harry Bennett,

David Fosdick, and Claire McKinney. Honorary pallbearers were H. W. Fox, C. J. Stauder, C. L. Browning, Dr. G. E. Clark, W. S. Davis and William Sanford. Flower bearers included Stella Pittenger, Ruth Munsell, Blanche Bennett, Tressie Knowlton, Reva McKinney, and Marie Nelson.

In the passing of William M. Black, the history of one of the West's colorful pioneers was closed. Mr. Black had lived in this district only since 1922, but the early years of his life in Idaho present a chapter in the building of the West that is almost without an equal in this year of 1951.

Mr. Black was born September 14, 1866, and the circumstances surrounding his birth were indeed unusual. His parents, Charles [Morgan] and Anice Black, were members of an oxen train en route from Chillicothe, Missouri, to Boise, Idaho. At a point some 16 miles west of Boise City the drivers one evening, knowing that Mrs. Black was about to give birth to a baby, stopped the wagon train and put the wagons in a circle in order to defend themselves against a possible Indian attack. Under those circumstances William Black was born. The Indians did come to the train and made it known that the wagons had to be moved on. They took one of Mr. Black's brothers and held him hostage and returned him a few hours later when the drivers promised to move their wagons ahead. The Indians made friends with Charles and Anice Black and allowed them to remain in that vicinity, but the rest of the wagon train had to continue onward to avoid a fight.

Mr. Black's father and mother decided to stay in that vicinity and they founded what was known as Black's Station, the elder Mr. Black operating a stage coach between Silver City and Boise for a period of 16 years. During these sixteen years young William Black played only with the Indian children, there being no other white children in that vicinity. He grew up, you might say, in the saddle, and from the time he was eight years old he was riding horses in the big races held at Boise on July 4th and other special days.

In 1882 William's parents moved to Wood River, where they engaged in the cattle business.

The country was all open at that time and the Blacks ran large numbers of cattle and furnished beef for the various mining towns in that locality. When the elder Mr. Black died, William took over the business.

Mr. Black was married to Etta Bellinger in 1893 and an unusual incident is connected. His bride was a member of the second wagon train from Oregon to come to the Wood River country. Four children were born to Mr. and Mrs. Black. They are Mrs. Fred (Violet) Griffith of Reno, Nev., Jeffery Black of Fowler, Charles Black, Huntington Beach, Calif., and Mary Wilson of San Francisco.

Mr. Black lived at Wood River until 1922, when he came to this locality with his son, Jeffery. They engaged in the livestock business on a partnership until 1930. Mr. Black was married in 1924 to Letitia Brunia, who along with two of her sisters homesteaded in the country southwest of Fowler. After his marriage he lived on the ranch his wife had homesteaded, and that was his home until he died.

Thus ended the life of a man who had known the rigors and hardships of early pioneer days and since lived to see and know the conveniences and marvels of our present day civilization.

Vignettes of My Mother

I was Mother's only child, although she claimed at one time that she had wanted a dozen children. During the Depression years having a large family may have had an advantage, but being the only child carries a great responsibility when all your mother's hopes and dreams are tied up in one "chick". She'd all but have died herself if something drastic had happened to me, and for my part I hoped to not be a disappointment to her in adult life. From grade school on there was never "if" you go to college, it was only "when" you go. She had high ambitions for me. She wanted me to be a politician or in some profession that could do good for the world.

Naming the animals was something Mother and I enjoyed doing together. Each spring a new crop of young calves came, and we selected a name for each. We could think of so many nice names, they never ran out. Chickens, ducks, and turkeys were given names only if they were special characters. For example, we named one of our chickens Little Yellow Hen because she was such a great mother. She would hide her nest of eggs and later surprise everyone by bringing forth a flock of chicks. She was skilled at maneuvering them among the snow drifts. Animals seem to recognize their name, looking up only when they hear their own name called.

Mother and I played Snap with old homemade cards she had brought from Iowa and played with as a child. I tried very hard to beat her by yelling "snap" first when two cards turned up alike, so you could take all the played cards. You won when you finally got all the cards. On rare occasions Mother would play solitaire, always smiling when she won. Some years later we got a Chinese checker board. It was hard to beat Mother until I gradually got wiser.

Mother had an old Victrola and several records from her early homestead days. Once in a while she would wind it up, put on a record, and place the needle arm on the record's outer edge to play. Miraculously, out would come Stephen Foster songs, Red Wing, Nellie Gray, and other old timers.

Mother, without doubt, had a deep, protective love for her child, one that I recognize more from my adult perspective than when I was a child. Mother did not have the best of teeth, plus she had painful bunions on both big toe joints. I was never to have either of these problems when I grew up—she made sure of that—and had begun before I was born by eating what would be good for the growth of her unborn child. She saw to it that I had milk for teeth and bones, even if she and Dad went without. When we didn't have a milk cow, I drank canned evaporated milk. My teeth, when they came in, were scrubbed with soda for five minutes, and to make certain they were clean, she did the brushing. When two of my upper permanent teeth came in crooked, she made arrangements in Pueblo to have them straightened. After each session, I was rewarded with a delicious strawberry soda. Somehow money was dug up to pay the dentist.

Mother developed bunions when she was a young lady by wearing pointed-toe shoes, which were the style. Going home after a party, her big toes hurt so badly she would take off her shoes and walk home in stocking feet. I would never get bunions, as my shoes were wide across the front, had at least a fingers width space in front of the toes, and high tops that supported my ankles. Once she bought me new shoes that lasted less than a week with my activity. She was very irritated that children's shoes could be made so cheaply.

Mother had a deep faith in God, His golden rule, the Lord's prayer, the Twenty-third Psalm, honesty, and fairness. Other than learning bedtime prayers, I didn't have any specific religious training other than the example of my parents. One glorious, warm, spring Colorado night, when the air was so clear it sparkled and the sky overhead was loaded with twinkling stars, we went for a nighttime walk. In this fabulous, open-air, exquisitely lit "church", designed beyond the makings of man, she told me God was everywhere, around us all the time yet invisible to our eyes. He could see us, she said, and understand our thoughts and the feelings in our hearts. He had created everything on earth and in the heavens—all the stars we see, the earth and all its plants and animals, the lovely pink phlox, meadowlarks, cedars, and hills. We could talk (pray) to God and ask for help when needed, and thank Him for our food, home, the cute baby animals, beautiful sunsets, and starlit nights. She explained that the most important of his rules was the Golden Rule, to treat others the way you want them to treat you. After a while we returned to the house. We never took another evening stroll and she never spoke much more about God. Mother had a way of telling things once, and somehow I knew it was to be remembered.

From the second grade reader Dad used to read me a story about a little black devil with short horns and a long, spear-pointed tail. I called him "the little black man". I enjoyed the story and often asked Dad to read it to me, but afterwards I was extremely scared of the dark when I had to go a ways from the house at night, and ran back lickety-split. Additionally, I had frightening nightmares. Mother had to figure out some way to cure this fear, since explaining he was not real did not make it stop. Eventually, she convinced me that we'd put him in a large wooden box and closed the lid so that he could not escape. It must have worked, as the nightmares ceased and I gradually again became fearless when going out alone in the dark.

One time I had a bad case of hiccups that just wouldn't stop, and normal cures such as drinking water or holding one's breath did not help. When I was not watching, Mother had got a box and put a chicken in it. When I came back, still with jolting hiccups, she said there were some books in the box, and handed me the box. Excited, I opened it, and instantly out flew a scared, cackling hen. I was so surprised that I didn't even notice that my hiccups were cured.

The Fowler dentist once made Mother extremely angry. I had a sore on a gum that didn't go away. The dentist said I had trench mouth and coated my mouth with a strong medicine. Very soon I got a horrible headache, the first one I ever had. The entire inside of my mouth burned like fire, and later became white. The lining became so seared that in a few days all this skin began to peel off. Naturally, it was a painful ordeal for me, but even more so for Mother agonizing over what had befallen her child. She raged and ranted over that terrible dentist for days. A few weeks after my mouth became bearable again, she took me to a dentist in Pueblo, who quickly found the problem to be simply caused by a toothbrush bristle lodged in my gum, which he pulled out.

Mother noted that I often breathed through my mouth. The wind seldom rested, and consequently there was always dust, especially in the corrals where the livestock stirred it up. Mother decided I might have a sinus problem and took me to a doctor in Pueblo, who without diagnosing, put tubes up my nostrils and pumped, and pumped, and pumped. I quickly got a tremendous headache and cried in agony, while my poor mother fumed in remorse. I scarcely could wait to get back home to the ranch. Perhaps just ambling around the arroyo would ease the pain. The treatment did not cure my problem, and Mother seethed every time she thought about it.

The mental agony of Mother always exceeded the pain of my injuries, possibly ten to one. Since we lived so far from school, for a while I boarded with another family that lived only a couple miles from the school. One night I was sitting at their kitchen table writing homework when the younger son came up behind me with a butcher knife. I was caught off guard with no chance to get out of the way. He struck for my face, brandishing the knife like a sword. To save my face, instinctively I grabbed for the knife receiving only a moderate gash on the temple, but my hand was something else, taking some weeks to heal. That weekend when I came home and Mother saw me bandaged up, she was simply beside herself with grief and anger, "Why you could have been scarred for life! I'll never have you away from home another year."

Mother was long-suffering with my cold feet. I liked to put them on her to warm, as in winter they were often cold. When the weather was severe, I was allowed to sleep in my parents' bed and put my cold feet on Mother's warm legs until mine were warmed and comfortable enough for me to fall asleep. Another big problem was that my feet often itched unbearably from chilblains, caused by exposure to cold and being frosted, producing an intense itching when the feet got warm. I begged Mother to rub my feet to relieve the stinging itches, which she did on many winter nights. Nothing could have felt more comforting. We did have a red rubber water bottle, and on cold nights she often filled it with warm water from the tea kettle and placed in the bottom of my bed. My parents had afforded two pink wool blankets for my bed, which comforted me until I left home.

I don't recall ever getting sick and throwing up, except for three times for the same reason. When Mother cooked for the WPA crew, she often cooked big batches of spaghetti. We'd never had spaghetti before, and I thought it exciting to zip up and swallow the lengthy spaghetti. A couple hours after I'd gone to sleep, I woke up sick to my stomach and threw up. Mother would hear me and come hold my head and offer me sips of water until I was all right again. Then she'd clean up the mess.

On the other hand, Mother dished out spanking and punishment when deserved. I was five the time I remember most. A Rhode Island hen had a flock of three-week old chicks with wing and tail feathers barely showing. I spent considerable time observing them when they were in the yard. I knew chickens could fly, so I began introducing the chicks to flying by tossing them in the air about two feet. With each unsuccessful attempt of flapping their stubby little wings, they plopped to the ground. From the house Mother noticed what I was doing and promptly came out and explained that the chicks did not have wing feathers and could not fly yet. When tossed in the air they could get injured by falling to the ground. "Stop. Don't do that anymore." I heard what she said and understood the reason. In spite of Mother's scolding, I was still tempted and amused by the chicks' struggle to fly. I did toss them more gently and not as high. When Mother looked out again and saw that I hadn't minded, without the slightest hesitation she grabbed me up and soundly paddled my bottom. Then she returned to whatever she had been doing. I hid out in the alfalfa patch and cried for some time, even though I knew it was fair and well deserved. She was a kind and gentle lady, but also very firm and strict. Whenever Mother explained the why for something and then told me not to do it, I was to mind the first time. There was never a second chance.

Mother advised me, as a child, on how to take care of my health and my body. She did say my body was beautiful once, so I would feel proud of the way I looked. I was to brush my teeth for five minutes, and when washing clean my face first and private areas last. After getting back from town we washed our hands thoroughly with soap in case of germs. One should go to the bathroom at the first urge and not hold it to become regular and to not absorb toxins. She also advised me on hair care and how to brush to bring oils down from the scalp. She taught me how to care for fingernails and to use an orange wood stick to keep the cuticle pushed back; and how to shape and trim my toenails to avoid ingrown toenails.

She taught me other practical lessons as well, such as airing the bed for thirty minutes before making it to allow moisture lost in sleep to evaporate from the bedding. I learned how to make a bed like nurses in a hospital with square corners so the bedding never slips. Many helpful bits of advice such as these have stuck with me all these years. She started me gardening as early as I was able with a small plot of my own, usually lettuce or radishes. She showed me how to make Hollyhock flower dolls, to darn socks, and to sew tiny cloth scraps together to make little quilts for my fairy dolls. She even wrote a poem for the Hollyhock flower dolls:

Hollyhock Ladies
by Letitia Black

Hollyhock ladies, demure and sweet
In ruffled bonnets of self-same hue
And fluted parasols shading you.
Dressed in gowns of satiny pink,
The prettiest dresses a child can think.
You are my lovely favorite dolls
In your frilly frocks and parasols.

Although money was normally spent frugally only for absolute necessities, Mother did not hesitate to spend on things she considered necessary for her child. When my teeth came in crooked, she took me to an orthodontist to have them straightened, so crooked teeth would not be an embarrassment to me in later life. Although homemade clothes had worked for grade and high school, once college bound, we drove to Pueblo to purchase whatever she thought other girls would be wearing. When I was getting married, she gave generously for the ranch wedding, purchasing a new couch to replace the old one (used since before I was born), a mirror to beautifully set off the mantel, and new clothes, including my wedding dress and shoes.

Mother had a decorated, glossy black, rectangular wooden box in which she kept some special treasures that I liked her to show me. One was a small china box about two by three inches in size. When holding this box to the light, you could see the loving faces of a grandmother and her grandchild, about to be bathed, etched in the bottom in astonishingly perfect detail. Another miraculous treasure was a tiny magnifying vessel, less than an inch tall, shaped like a telescope. If you peered in the tiny opening at the top while holding it to the light, you could read the Lord's Prayer. How could anyone have written it so small? These treasures were probably crafted in the 1800s or before, long before my time. She also kept the three tiny arrowheads, some old-time songs written on brown paper, and a few pieces of jewelry in her decorated treasure box.

Idle never existed in our vocabulary. Mother started me on many hobbies to help me entertain myself while learning. I had quite a solid beginning of a stamp collection. Uncle Will sent a huge stamp album with illustrations of all the US stamps up to that time. There was the button collection with several boxes, no two alike. We were always on watch to find new stamps or unusual buttons. My beautiful butterfly and moth collection eventually got ruined by clothes moths. Mother walked with me to various habitats, such as the canyon, where certain butterflies lived. She made a net, and we'd catch and carefully bring home butterflies to add to the collection. I also collected cocoons and pupae to hatch and watch their metamorphosis. Huge green spiked tomato worms found on the woodbine by the house would turn into lovely lavender sphinx moths, and the stripped worms on the dill became tiger butterflies. Another hobby was making scrapbooks. Aunt Barbara, the lady Mother named me after, started this hobby by sending me two scrapbooks.

When I was older, Mother started me with 4-H projects and food preserving, mostly canning and drying. This was a good experience for me to meet other youngsters my age. Dried apples had to be flexible and leathery, not hard and brittle. My canning skills were good except for canning meat, especially rooster. I won red and blue ribbons at the county fair in Rocky Ford, and because of winning there I got to exhibit at the state fair in Pueblo, where I won a purple champion ribbon.

During my college years, Mother feared the cafeteria meals might not be nourishing enough so she bought a metal egg container holding several dozen for me to take each time I came home. As a child she had explained to me that an egg contained all the nutrients needed for the development of a chick; its bones, blood, flesh, everything, and likewise was nutritious for us. (I soon got a hotplate and did cooking in my dormitory room.)

My mother had very beautiful, easy to read handwriting. She kept up correspondence with the Iowa relatives, her nursing friends, the homesteaders that had moved away, and many other friends as well as business affairs. She kept all the accounts and perfect records of all expenses and income.

Mother loved to read and considered the National Geographic magazine important, never allowing her subscription to lapse. (By the time of this writing, we must have 80 years of National Geographics.) Sometimes, in the afternoon, if the work was under control, she would sit and read Good Housekeeping, and later also afforded The Saturday Evening Post. She'd sip a cup of hot tea and munch a cookie as she turned the pages, absorbed in the articles. There would be continued stories, so she had to wait until next month's issue to continue, for they were sure to leave off at a place of suspense. The other time Mother read was at night by the kerosene lamp, generally peeling and quartering apples to eat as she read, sharing pieces of apple with Dad and me. Some winter nights she read us stories from her twelve volume set of The Stoddard Library. She also had volume sets of Shakespeare and Kipling. During the winter evenings, I remember the comfort of the kitchen stove while Mother sat at the table and read stories to Dad and me by the light of a kerosene lamp. From her volumes of great literature came stories, such as the Arabian Nights, that held me spellbound.

When we were away from the house, riding or walking doing ranch work, we never took water as we all seemed to be good "camels", drinking before we left and when we came back. Water would have been just one more weight to carry. Mother did take a small empty vanilla bottle with water in her pocket on a hot day, mostly just to wet her mouth.

Whenever we were out near the canyon with a car, Mother found various sized slabs of sandstone to bring home for building a rock wall between the house and the yard gate and continuing from the gate toward the corral. She was expert at fitting these big, heavy rock slabs together, often turning one carefully several times until it fit perfectly. Eventually, she created a graceful, durable wall two feet wide and

By the canyon gathering rocks for the rock wall.

four feet high with a slight, gentle curve. Not only was the rock wall attractive, receiving many compliments, but it was protective against the ever-blowing wind and dust entering the house when the door was opened.

I have vivid memories of three times that two or three cars of Gypsies drove onto the ranch, crammed to the gills with parents and children of all ages. One of those times Dad said he lost the billfold out of his back pocket and never felt it being taken. Mother had heard stories of Gypsies taking children and wanted to keep an eye on me, but she was torn between that and wanting to watch what food the mothers were helping themselves to. One Gypsy lady in her full and colorful skirts wanted to go down into the cellar with Mother to get potatoes, saying "I'll help you." Mother was frantic because the lady was grabbing handfuls of potatoes right and left, and Mother couldn't afford to spare so many. While she was in the cellar, she had no idea what the others might be taking. The big groups were like a whirlwind of activity. I didn't share my parents dilemma, being entranced by all the children and clothes more colorful than I'd ever seen.

A year came when the folks decided it was time to have electricity. They got a wind charger to place on top of the house and strung out an aerial line just above and to the south side of the house. They put up some 12 volt lights and bought a radio. This project was short lived, as a strong wind blew the wind charger down, bending its central metal pipe in a "U". While it lasted, Mother enjoyed a few mid-morning radio programs and music. One commentator said, "Make yourself an honest man, and then you will know there is one less rascal in this world." She liked that, and she liked other commentators of the time like H. V. Kaltenborn, Eric Sevareid, Edward R. Morrow, Paul Harvey, Lowell Thomas, and Walter Winchell. She also listened to ranch programs and music. Some years later they purchased a generator and batteries for electricity.

When it was time for bed, Mother removed the pins holding her hair in a bun at the back of her head and brushed her long brown hair. Then she cleaned her face with Lady Ester cold cream, which helped keep her skin soft in spite of the harsh outside weather and sun. Lastly, she rubbed a little sheep tallow on her rough hands and cracked fingers. When Dad and I had turned in, she checked that the stove was closed down, turned the lamp low before blowing it out, and she, too, went to bed.

Epilogue

I ended the story of my mother's Colorado years with June 1950, after I had graduated from college and Mother and Jeffery came to bring me home. Afterward I got married and left the ranch (see Eulogy to Bill Henderson). Dad lived another year and two months until August 1951, and near the last, he required much care from Mother, leaving her worn out.

Cousin Jeffery Black, when we were 21.

Mother continued to manage the ranch by herself for another fourteen years, running about 100 head of cattle, a number suited to her 4000 acres, or approximately forty areas per head. She also kept a few saddle horses. During these years she persevered through two broken arms at once, and later a broken hip. She faced difficult winters with frozen water to thaw, cattle to feed, and the ongoing care of livestock and ranch maintenance.

In the fall of 1964, with my encouragement, she sold her cattle and moved to the lovely Sonoma foothills outside of Healdsburg, California, taking the last remaining horse, Black Bess. With the sale of the cattle, she was able to purchase a small acreage adjoining our home property to be near us and enjoy her three young grandchildren: Tamara, Roxana, and Keven. My cousin's family, Jeffery and Ilene Black, then leased the 3N Ranch, where they raised their three children: Bill, Tim, and Lynette, all horse lovers, and ran cattle and saddle horses.

Mother in California with grandchildren Keven, Tamara, and Roxana.

Mother had an acre of Zinfandel grapes, a few prune and pear trees, a shed for chickens, and a small garden plot. She pruned and harvested her grapes, selling them to a local winery. She enjoyed her grandchildren stopping by, and always had baked treats. But this brief, relatively carefree period of her life did not last long, as in the fall of 1968 she suffered a serious stroke from which she did not recover enough to care for herself or her place. Not only did she need help to walk, but she could no longer swallow and speak. Her food needed to be pureed and drinks given to her by spoon. We added a room to our house, Letitia's Lookout, with windows all around for her to live with us. In May 1974, we moved to Happy Camp, California, again preparing her a room with spacious windows.

Dear Mother, she never complained and never changed in her sincere demeanor. She could communicate with us by writing notes, and she often asked for something to do by holding up her hands to indicate they needed work. Every neighbor that met Mother, highly complimented this smiling, cheerful woman, wishing she could speak as they visited.

Mother's Farewell Banquet

Mother always served meals like a banquet for hungry ranch hands, an ample variety of food in large quantities. This was not just true in times of plenty, it was always true. She often did not know herself where the next meal was coming from during the Depression years, when their last ten dollar bill was tucked in the Bible, or during the long drought, when her prairie grassland turned from green to dust. Yet, miraculously, three square meals a day appeared on our table, and even more sumptuous meals if guests showed up.

But the dinner I want to tell you about happened the day after Mother died. She died at home on our "Nine Springs Ranch" in Happy Camp, California, a little after nine pm on August 16, 1974. I stayed up that night to be with her soul and body as they were adjusting to new dimensions. I, too, was adjusting, my thoughts muted as I bathed her. There came even a hint of relaxing happiness while choosing her favorite, most-becoming garments, pinned a cameo, then located the maroon, marble-sized beads she used to wear to town when I was a child, as I dressed her that last time. She looked so nice.

About four in the morning I matter-of-factually said to Mother, "Oh, what will we do for meat tomorrow?" Although she breathed no more, I was so used to having a mother all my life, I forgot she could no longer answer. All the children and their families were coming, and I wanted to have a nice meal for them. Suddenly, I had remembered that both our checking account and billfolds were empty.

Mother didn't take two seconds to respond in the wordless way she now must communicate in her newly born afterlife. After a tiring, seven difficult years in a stroke-afflicted body, her answer came in the twinkling of an eye. Already her decision was sharp, clear, and quick, for "bang! crash!", a deer ran into the fence scarcely twenty feet outside her bedroom window. But more amazing, the deer stayed stuck long enough for me to streak out to the barn, grab a rope, and dash back to get the rope tied around its thrashing neck. Still afraid it might escape, I ran for a second rope to secure it even tighter. The deer bellowed, shrilled, snorted, and made other piercing sounds I had never heard from a deer before while struggling and plunging wildly. But it remained tied and stuck until my husband, aroused by such a commotion, ran out with gun in hand. I am certain that Mother thought a fine venison roast would be just the meat for dinner. How else can a strong, panic-stricken deer staying entangled longer than a split second be explained. It could have sailed effortlessly over the mere four-foot fence, or just as easily burst through the wire leaving behind only its hole for us to mend.

The next day our children's families arrived (except for Keven, who was still living at home with us), including Mother's grandchildren and her first great granddaughter, Valerie. We were all busy preparing for her service the following day, each writing down the things they wished to say, the little things she would appreciate knowing had been remembered. We invited a half dozen of the neighbors she knew to come to her service. Plots had been purchased in the Happy Camp cemetery, and in the afternoon we took shovels and picks to dig her grave. Our family always dug and prepared the graves for our loved ones and close friends as the last parting gesture we could do for them. That evening we were all hungry and enjoyed the delicious venison roast that Mother had "planned" for our dinner.

Since Mother had always loved apples and apple trees, her service was held in the front yard under an apple tree. Mother had chosen red roses and shafts of wheat for Dad's casket. We did the same for Mother's, as a red rose was blooming by the clothesline and there were shafts of wheat growing in the big garden. It was a typical hot August morning without a cloud in the sky when the hearse arrived with Mother in the casket. We walked out to receive Mother back when I felt big rain drops hitting me. Very surprised, I looked up to the sky and there was a big black cloud overhead showering rain on us. I took it to mean that even the heavens were shedding tears in big drops for this wonderful lady. The casket was carried under the apple tree, and I checked Mother over to be sure she was properly dressed again the way she had been when she was taken away for an autopsy. Then I glanced up to inspect the black rain cloud. I couldn't believe it had vanished. There was no trace of it, only blue sky again. The neighbors came, we had her service including the Lord's Prayer and the Twenty-third Psalm she liked, and then we saw her off at the cemetery.

The next day the children and their families left. It was unusual around the house, as if Mother were there and yet she wasn't in her room. Kind of an empty or lost feeling. We had a fourteen-year-old small chicken, Lucy. In her old age Lucy spent much of the time in the house, and especially in Mother's room. It was nice and sunny there. Some chicken hens sing more than others, but Lucy out sang any chicken we ever had. She could sing for an hour non-stop, which she often performed by Mother's bed. Lucy always made Mother smile. This day Lucy didn't sing for an hour; she sang at the top of her lungs all day! I'll never forget that day. The day after that Lucy did not sing ever again. She was her normal self, except she had silenced her cheerful voice.

One of the local ranchers was also an artist of ranch life and a sculptor. We asked him to sculpt a bronze calf as the headstone for Mother's grave. This young man had qualities similar to Mother, a real rancher at heart, combined with the talents of an artist. Mother would deeply appreciate the calf he made to stay with her.

Her headstone below the calf reads:

"She loved the new spring calves; valued neighbors,
prairie flowers, sunsets, clear starry nights.
She was skilled with the wood cook stove,
had apple pie for lonely cowhands,
left a lamp lit on blizzardy nights should anyone be lost.
Every challenge of ranch life she faced."

Appendix I: Ranch Folk Sayings

The folks had a number of saying they used in everyday conversation. Here are ones I still remember hearing.

"Where there is a will, there is a way." Mother said this for something difficult to tackle.

"Till the cows come home." You might have to wait a long time.

"Go to bed with the chickens." Go to bed early, like the chickens.

"Run down in the heel." Referring to something worn out, like old shoes.

"Strike while the iron is hot." Make use of this opportunity now.

"Hit the nail on the head." You said it exactly right.

"A stitch in time saves nine." Fix a problem now and it will save a lot of time later.

"Thread bare." Thin, worn out (said of cloth or clothing); poor or shabby; ineffective.

"Worn to a frazzle." You've been working too hard or worrying too much.

"Do one's level best." To do the best one is capable of doing.

"Won't amount to a hill of beans." Not contributing anything of value to life.

"Would give the shirt off your back." Meaning a very generous person.

"Bit off more than you can chew." You attempted something beyond your capacity.

"By the skin of your teeth." You just barely made it.

"Don't let the grass grow under your feet." Get a move on, don't waste time.

"Fell into your lap." Said of something desirable received without effort.

"Let the chips fall where they may." Said about possible outcomes you can't control.

"When our ship comes in." When we make our fortune.

"Gone to pot." Not kept up; gone downhill.

"Three square meals." Having three substantial, nourishing meals each day.

"Eyes bigger than your stomach." Putting more food on your plate than you can eat.

"You can't have your cake and eat it too." You can't have two contradictory things at the same time.

"Child, you haven't got the sense you were born with." Said if I did something stupid.

"Don't let it get your goat." Don't let it bother you. (Goats can be very annoying at times.)

"Stubborn as a mule." You hate to give in and want your own way.

"Has the cat got your tongue?" When someone can't think of what to say in response.

"Don't let the cat out of the bag." Don't tell anyone. Keep it secret.

"Don't spill the beans." Another way of saying, "Don't tell anyone."

"Take a spit bath." Taking a sponge bath with just a tiny bit of water.

"Make hay while the sun shines." Take advantage of an opportunity while conditions are good.

"A penny saved is a penny earned." Money not spent is money in your pocket.

"Save your pennies and the dollars will come." The folks said this sometimes.

"Scrape the bottom of the barrel." Having to use the worst or poorest available. Used to the last bit.

"At the end of the rope." Out of options.

"You can't get blood out of a turnip." You can't get from people what they are unable to give.

"He's a good egg." He's a good person.

"Closed the barn door after the horse has bolted." You tried to fix something after it's too late.

"Fly off the handle." Losing one's temper suddenly and unexpectedly; also said when a saddle horse suddenly spooks and bolts.

"Keep your nose to the grindstone." Keep busy and don't get distracted.

"Battle through it." Mother said this when we had a difficult job to do.

"Step it up." Mother said this if I was poking along behind too slow.

If something shouldn't take too long to do, it will get done "in a jiffy" or "in a whip-stitch".

When going too slowly, taking it easy, Mother would say "pottering around" or "dawdling".

Mother liked people that had "gumption" and "common sense".

"Knee high to a grasshopper." Very short or very young.

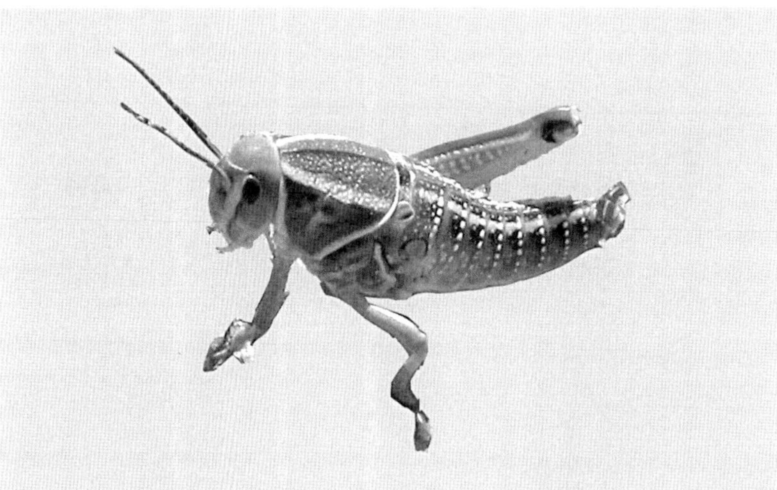

Lubber grasshopper from a draw near Haystack Butte.

Mother's colander, flour sifter, bread crock, and measuring cup.

Appendix II: Mother's Recipes

Unless Mother was making something new, she rarely ever used a recipe; hence few of her recipes were written down, except for those that I wanted later for my own home. Also she seldom used measuring cups or spoons, adding ingredients mostly by guess, a pinch of this or a dab of that. Recipes were often varied or adapted by what food or ingredients we had on hand, or a taste test before being cooked. Everything Mother made was delicious. She was a culinary artist.

Sour Cream Cake

Cream together:	**Sift together:**
1½ cup sour cream	2 cup flour
1 cup sugar	1 tsp salt,
2 eggs	1 tsp soda,
1 tsp vanilla	1 tsp baking powder

Bake 30 minutes at 350.

To make a chocolate sour cream cake (which Mother generally did), add 4 tbsp cocoa and a little less flour. No baking powder, but 2 tsp of soda in a little boiling water added at the end.

Chocolate Cake

½ cup butter	3 tbsp cocoa
1 cup sugar	1 tsp baking soda
1 egg	1 cup buttermilk
1½ cup sifted flour	1 tsp vanilla

Cream together the butter and sugar until fluffy. Add beaten egg. Sift together dry ingredients. Add buttermilk and vanilla. Bake at 350 degrees about 35 minutes.

Seven Minute White Icing

1 cup sugar	1 egg white
3 tbsp water	¼ tsp cream of tartar

Beat with eggbeater, cooking in double boiler over rapidly boiling water for 7 minutes (4 minutes with an electric beater) or until icing stands in stiff peaks.

Fruit Cake

½ cup shortening
1 cup sugar
2 eggs
1 cup sour buttermilk
2 cups flour
1 tsp cinnamon
½ tsp allspice
¼ tsp cloves
1 tsp baking soda
1 tsp baking powder
1 cup raisins
1 cup watermelon preserves
(Mother made our own)
dried orange peel pieces

Cream together the shortening and sugar. Add beaten egg and buttermilk. Sift together the dry ingredients. Combine together and add other ingredients. Bake at 325 degrees for about 1 hour and 30 minutes.

Molasses Ginger Cake

2½ cups sifted flour
1 tsp baking soda
1½ tsp powdered ginger
½ tsp salt
¾ cup currants or raisins
1 egg
1 cup dark molasses
1 cup buttermilk
¼ cup melted butter

Sift together the dry ingredients. Beat the egg with molasses and buttermilk until well-blended. Add butter and mix in dry ingredients. Turn the batter into a greased and floured 9″x5″x2¾″ loaf pan. Bake at 325 degrees for about 1 hour and 10 minutes. Let stand 5 minutes before turning it out of the pan. Let it cool down, then add frosting if you like.

Lemon Meringue Pie

1 cup sugar
5 tbsp cornstarch
1½ cup boiling water
1 tsp butter
3 egg yolks
juice of one lemon and rind
3 egg whites
1 tbsp powdered sugar

Sift together the sugar and cornstarch. Dissolve it in boiling water. Stir in butter, grated lemon rind, egg yolks, and lemon juice. Put it in a double boiler and cook it until it turns slightly transparent. Pour into the pie crust and let cool. Whip the egg whites. When almost stiff, gradually add powdered sugar. Beat until stiff peaks form. After the pie filling cools, add the meringue. Bake at 325 degrees for about 20 minutes.

Oatmeal Cookies

1 cup sugar
1 cup shortening
2 eggs
½ cup sour milk
2 cups flour

1 tsp cinnamon
½ tsp salt
1 tsp baking soda
1 tsp baking powder
1 tsp vanilla

Cream together the sugar and shortening. Stir in eggs and milk. Sift together the dry ingredients. Mix all together. Add 2 cups oatmeal, 1 cup raisins, and 1 cup nuts. Form into cookie-sized balls on a baking sheet. Bake at 360 degrees for about 10-12 minutes.

Santa Claus Cookies

3 cups flour
1½ cups brown sugar
1¼ cups butter
¼ tsp cloves
½ tsp allspice

½ tsp cinnamon
2 eggs
2 tbsp water
1 tsp baking soda

Mix the flour, sugar, butter, and spices. Then add eggs and water in which the baking soda has been dissolved. Nuts may be added. Bake at 350 degrees for 8-10 minutes.

Upside-down Cake

4 tbsp butter
1 cup brown sugar
□ cup shortening
□ cup sugar
2 eggs separated

□ cup buttermilk
1½ cups flour
2 tsp baking powder
1 tsp soda, ½ tsp salt
½ tsp vanilla

Cream together the butter and brown sugar. Cook on low heat in a skillet until smooth, and syrupy. Remove from heat. Arrange well drained cooked pitted prunes, pineapple, or other fruits in the syrup.

For the cake, add sugar slowly to the shortening. Add beaten egg yolks and vanilla. Sift together flour, baking powder, soda, and salt. Combine mixtures adding buttermilk, a little at a time. Lightly fold in beaten egg whites. Pour the batter over the fruit in a skillet. Bake at 350 degrees for about 45 minutes. Remove the cake at once and turn it over so the fruit is on top.

A couple arriving to visit the Three Nurses. If not in a buggy, the other way visitors came was either horseback or in a wagon.

The old abandoned homesteads always intrigued me. I contemplated who had lived there, such as in the old log house which was past the canyon, on up the draw of Saunders Arroyo around a bend just before Cactus Flat starts in earnest. By some miracle it lasted many years before a cyclonic wind finally took it down.

Appendix III: Homesteaders and Neighbors

These are some of Mother's fellow homesteaders, neighbors, and cowboys that dropped by, and other ranch owners within a radius of some twenty miles. Most have passed on except for a few from the younger generations. In 1974, before they passed on, Mother's sister, Henriette, and our neighbor Blanche Bennett, put together a few notes about the years 1916 to 1924.

In the homestead days, when people lived miles apart, telephone lines didn't exist, and transportation was mostly on a saddle horse or by team and wagon, neighbors felt a strong bond and loyalty toward each other. They experienced the meaning of friendship, the neighborliness of helping out, or just dropping by to say hello and inquire about everyone in the family. When company came, Mother would stoke up the wood cook stove and have chocolate cake or pie baking in minutes as they talked. Neighbors never left without sitting to a cup of tea or coffee and baking still hot from the oven. Sharing troubles and trials was a means of shedding stress, and if help was needed, it was offered without hesitation.

August and Gladys Neihaus (daughters Nadine, Helen, and Jeannie) came most years to visit and see their homestead, already blown away by the winds except for the cistern and an old iron wood cook stove. Their homestead joined Aunt Henriette's on the north, Red Top to the west, and was otherwise surrounded by the Sanford land. Mother and Gladys would rehash the early days and wrote frequently to each other. As their land had no water, Gladys told how she drove their milk cows several miles through range cattle to the Cross Bar to water, where she filled two jugs to carry drinking water back to their homestead. She was known to shoot prairie dogs for their supper. She wrote that a neighbor homesteader once went out in a snow storm when his hat blew off. When he tried to retrieve it, he got lost and froze to death. She rode one of their work horses 14 miles to get the mail, which was delivered twice a week. Their milk cow was rattlesnake bitten, which she cured with kerosene, but she got sick when drinking the milk.

Frank and Myrtle Leonard had homesteaded south of present Highway 10, close to the Sanford Hogback. They came from Kansas yearly to see the land where their old place once stood, visit, and stay overnight. They always brought "their friend" (a commode). Mother would put out nice things from the trunk to make their bed. They wrote to Mother frequently. One time they showed up unexpectedly with Blanche Bennett, who said, "We came for dinner, Letitia." She replied, "Wait til I get a chicken and dress it." With a wire hook she caught a young rooster by the wood pile, scalded, picked, and cut it up for the skillet, then got vegetables from the garden, and made the best rhubarb pie they'd ever tasted, all while they visited.

Leo and Elsie Carr had a white frame house on Saunders Arroyo just north of Highway 10, which withstood the elements for many years; always an interesting place to explore when out riding. Several of their family had homesteaded in the same area, making sufficient acreage for raising cattle and running a number of horses. Leo was quite a horse trader. He kept a good work team and a pair of fast trotters for the buggy, which Elsie took to Fowler to get supplies. As a young girl, Elsie could have any horse she could break, and rode spirited horses to teach school and to visit the three nurses. I asked Elsie if she could harness a horse. "Oh sure, as fast as any man, was raised that way." Their chicken flock was securely locked up nightly against the ever lurking coyotes and other predators. Having a good well, they always had big gardens. Elsie, and later their daughter, Pauline, corresponded with Mother.

Harry and Nella (a sister of Leo Carr) **Fassett** lived west of the Omar School near the Mustang and raised cattle. Nella loved horses and had a favorite cutting horse named Polly. The Fassetts were a big help to the three nurses when they started their homesteads. They let them stay with them while Letitia's cabin was being built, and later sold the sisters some cattle and a riding horse. Mother kept in touch with the Fassetts, and we once visited Nella in Durango. The Fassetts were the family that originally inspired the three nurses to homestead.

Fred Jones had a homestead that joined ours to the north by the Jones Reservoir, where migrating ducks and geese would stop over in fall and spring. Fred just had a piano box as his cabin and plowed his forty acres to "prove up". He worked in Pueblo, and when away from his homestead, he left a pair of his overalls hanging on a line by his "cabin". Mother later bought his land with her turkey money.

The Mobly homestead joined Mother's on the south. She purchased their section of attractive land with buttes and a limestone ridge running north and south with cedar trees. It is located just north of the Woods pasture and west of Duffy Crow's homestead. There is a half mile slopping break in the ridge, and their log cabin was built in the valley between. I always had to peek inside when riding nearby. It stood for many years before finally being demolished by a whirlwind. In later years, Mother had a reservoir put in the draw just below where the cabin had been.

Duffy Crow's homestead was at the end of the Sanford Hills, north of the school section, and it joined the Mobly on the east. Exploring his dug-out dwelling was exciting for me seeing parts of his old wagon still there. Once when Vess Petry was caught in a blizzard, he made a dash for the safety of Duffy's dug-out. Vess was good friends with Will Barry, who homesteaded just to the west of the Mobly.

Will Barry's section joined Mother's to the southwest. She purchased it from him after he moved and married her sister Henriette. In prior years, drilling on his land was done in search of oil, and two big wooden cable spools, strands of cable, and other remnants of the drilling were left behind. He had dug a well by Saunders Arroyo, but the water, like that at our place, was strongly alkaline.

Lou and Marie Nelson, and their daughter Mary Lou, lived in a rock cabin up a canyon east of the Flannery and south of us. The Lily Pond was on their land, where they ran some cattle. In the early years, the wagon road from the south to Fowler went through our barnyard, as Saunders Arroyo was shallow enough there on both banks to ford it with a car or wagon. The Nelsons would stop briefly to say "hello" to us. Lou had been in the war as a young man.

John Flannery lived in a partial dug-out some four miles to the south of us, and he had a couple of work horses, Old Bill and Old Silver, and colt named Little Billy. He drove his team through our barnyard about once every couple months (less often in winter) on his way to Fowler for supplies. He never said a word, just a grunt to acknowledge our "hello", so we really never got to know him. In two or three days he'd pass back through our place, his wagon loaded with supplies. There was a reservoir with a rock dam on his place, which with summer thunder storms held water for his horses. After he died and a son came to get what he wanted, we found many nice and interesting items at his place. One was a lovely half gallon china pitcher we used for making lemonade. He had a brother, Jim, who homesteaded a mile east of him and lived in a square frame house.

Jim and Ida Bell Sanford (children: Alone, Mary, Alma, Carry, and half brother Harry Noble) and his brother Bill homesteaded together on what was known as the "Bill English". The family lived in a several room cabin located on the fairly steep hillside just below the ridge of the Sanford hills during the summers for the first three years required for "proving up". Bill, younger than Jim by four years, never married and lived to 100 riding hunched over at cattle round-ups. Harry Noble's homestead was between our horse pasture and the Sanford Hills. The Sanfords had acquired thousands of acres as other homesteaders moved on, and they ran hundreds of Hereford cattle. Saunders Arroyo ran through their land, having seepage where their stock could water. The Sanford Hills and Buttes, and the Hogback to the east, were all on their land. They also owned the Blue Ridge reservoir on the west side of their property. Alma was a great part of my childhood and gave me one book each year for Christmas according to my age level.

Photo of the "Bill English" old homestead cabin. Nestled tightly against the steep hillside just below the crest of the Sanford Hills, it has withstood the winds that have destroyed the other homestead cabins. After the Sanford family, it was home to their cowpuncher, Bill Henderson, from 1927 to 1972. It had a water cistern and hitching post in front, a woodbine at the northwest corner, a big kitchen living room, and three bedrooms, all with exciting things, and nicely kept.

Photo courtesy of Jim Cash, grandson and namesake of Jim Sanford.

Anthrax is a dreaded cattle disease that I heard about as a child. By great fortune it never infected any of our cattle, but it did infect some of the Sanford cattle as well as those of the Krieders and the Pittengers. It primarily occurred in drought years, when cattle cropped grass close to the ground, picking up anthrax spores from the soil, predominantly in alkaline areas, and would drop dead within three hours. Any animal that died from it had to be buried deep or burned immediately before buzzards or coyotes tore open the carcass and allowed anthrax spores, which can survive for decades, to seep into the soil. People were not to touch the animal.

Floyd and Eva (Leap) Krieder (children: Ila May, Warren, Billy (wife Winnie), Irvin known as Curly (wife Ella May and son Carl), Frances, and Della Lee) lived in White Rock. Saunders Arroyo runs through their land. At three Billy set their house on fire and was made to wear a dress as punishment. When young, Della had climbed their 40 foot windmill tower. Her mother told her to come down and "I'll tan your pants!" Della Lee replied, "Then I won't come down." Once Eva and her girls visited a neighbor with a new bathroom. It was so much nicer than their outhouse that Della Lee had to go in every ten minutes. What impressed me at their home was that their wood cook stove had a reservoir for hot water. The Krieders ran white face cattle and saddle horses on their land joining the Sanford ranch to the west. Floyd at one time cut cedar posts, hauled them to Fowler to sell, and brought back the mail to the White Rock post office (he got paid $1 a trip).

Herbert and Tressie Knowlton (children: Allen, Harry (who died of polio in grade school), Ralph, and Nadine) lived in White Rock, where they homesteaded on Saunders Arroyo. Tressie wrote that they lived in a dugout the first year they were married. She would see Henriette Brunia, one of the three nurses sisters, riding horseback through their pasture on her way to teach at the White Rock school. They dug a well for water, and afterward built a big, white two story house, always had a garden, and ran cattle and horses. I went to school with their children, who were talented with music and art. Before I started school, when the family came to visit, the boys were the only neighbor children I knew and got to play with. Mother and Tressie laughed and giggled like two young girls when they got together. Herbert was an engineer. He had surveyed the Apishapa dam and all the local irrigation ditches as well as elsewhere.

In later years, **Allen Knowlton** married **Melba Rafferty**, and they lived five miles west of our ranch below the Hogback. They had a sizable family, and one of their sons, Bryce, and his wife Ronda, leased our 3N ranch in 1994. (Allen had leased it too in 1978 after Jeffery's family moved out.) Their daughters Addie and Tressie were talented horsewomen, and both won the title of Miss Rodeo Colorado in different years. The girls set up a sizable area between the yard and the arroyo with barrels to practice and train their horses. Tressie went on to win the title of Miss Rodeo America in 2006. We flew to Las Vegas to see her win the crown. It was a great and proud moment of my life, knowing that Tressie had been raised on our 3N Ranch.

Nadine Knowlton married **Cletus Martin**, and they live on the original Knowlton homestead. Cletus has the equipment for working on wells and maintaining and repairing windmills. This is very important in this land where rainfall cannot be counted on to keep water in the stock ponds.

W. S. Davis lived in a big two-story house in White Rock. He was not a rancher, but lived on his wife's $125.00 pension. He wore a mustache and drove a black coupe to pick up his mail. He had four daughters: Ellen, Alice, Edith, and Isabel; and two sons: Algernon and Dannon. Algernon, who had gone to college, died by suicide when rabbit hunting. Dannon, who had been in WW I and afterwards come home and married, sent his wife and baby girl Libby back to Kansas to live with her mother. On their table was an unusually large crystal bowl with ear pieces. After his first wife died of asthma, Mr. Davis married a lady named Leo by mail order. She was an educated woman and a musician among other things. After several years she left.

Merle and Arlene Black (children: Bud, Jim, Bob, Beth, and Lucy with red hair) lived close to the Davis's place on White Rock Road. The three boys rode to school on a tall bay horse. With no well on their land, they had to haul their water from a sulfur spring a mile away in the Krieder's pasture. After they moved to Fowler, Jeff and Sara Black rented their place, which had originally been the homestead of George Eddy.

Elton Watson and Lucille Huntley went together for a few years before they married and lived in the old two-story Molholm place. They had one son, whose wife took care of Elton after the son died. Elton's father-in-law, Floyd Huntley, had been a meat cutter in Fowler. Lucille used to go on the White Rock school picnics, bringing homemade salad and cookies. Her brother Paul Huntley would play his fiddle at the White Rock dances.

Jason (Snake) Blakeslee lived at the junctions of South, Bass, and Apishapa canyons.

Vess Petry and his wife had a place near the bottom of a canyon south of the original Pittenger homesteads with their horse pasture extending into the Hall land. Their daughter, Carolyn, married Clark Harriman, and their son, Harold, worked on the Red Top Ranch. Lover's Cave was in the canyon near their place, and it has dripping springs, where the homestead ladies kept tubs to collect water and an old stove to heat the water for laundry.

Harry and Blanche Bennett (and daughter Betty) lived south of the Sanford foothills. Their place and the Hall's were originally homesteaded in 1905 by Blanche's parents, Robert and Mable Pittenger. Mr. Pittenger had been the foreman of a big ranch in Texas. Blanche was three years old when the family came to Colorado in a covered wagon (making 10 to 12 miles a day) bringing a herd of cattle along with them to the homestead. The deed to their land was signed by President Woodrow Wilson. Robert Pittenger later divided his land between his children: Fred, Blanche, Cora, and Mary Denny. The Apishapa Dam, started near their property in 1920 and completed in 1922, was built mostly by four-horse teams pulling wagons equipped with tipplers to unload their loads of dirt and rock. They also pulled the water wagons and heavy rollers used to compact the earth. Blanche cooked for nearly two hundred men during that construction project. When the dam broke and washed out in 1923, it cut a four foot gully in their land. Blanche later cooked for the pipeline workers. She was always thoughtful of me, giving me a small present at Christmas time in the name of Santa Claus, not wanting it known it came from her. She passed on some of Betty's out-grown clothes to me as well.

Harry and Cora Hall (and sons Ray and Roger) lived a mile west of the Bennetts along Buffalo Arroyo, where an old time trail existed along which cattle were often driven. Cora and Blanche were sisters. The Halls raised chickens. Roger, who married a girl named Dora, once shot an antelope and told us it was a sheep. It was cooked in a pit with rocks on top and was delicious. When Ray was small, he accidentally started a fire, and they lost everything. When the Apishapa dam broke, their fences and water gaps were washed out. The flood left so much deep mud in low places that their cattle often got stuck and had to be drug out. Once they tried to pull a cow out by the horns with a lariat, but its horns broke off and the cow died.

Lois and Henry Hamilton lived east of the Pipeline and managed a cattle ranch. Dorothy Mumm, a relative, stayed with them a while and attended White Rock school. Lois often went on our school picnics and brought nice things to eat.

Ruth and Forest Munsell (children: Howard and Yvonne) homesteaded south of the Apishapa, a very lovely spot with scattered pinyon and juniper. They ran some cattle and a few horses. Ruth later lived in town for the children to go to school. They were like family to me. Howard did leather work, even making saddles. Yvonne was two days younger than me, and I considered her my best friend. Ruth was like a second mother, famous for her divinity candy.

Fred and Stella Pittenger (and sons Wayne, who had epilepsy, and Buck) lived in Fowler and always had a beautiful weeded vegetable garden. Stella and Ruth Munsell were sisters. Ora White had a duplex home just west of the Pittengers, where I stayed to go to high school for two years until she died from shingles. She had been a homesteader and was Ruth and Stella's mother.

The Arnold/Harriman Ranch, where **Clark and Carolyn, Frank and Ed Harriman** lived and ran sheep, was on both sides of Highway 10 east of the Apishapa River and had many sections. Acres of yucca grew near Highway 10 on their land. Vinson and I got bunches of yucca flowers from there to decorate for our wedding.

The Ingram family lived seven miles west of Fowler, where they raised sheep and farmed by Dotson Lake. I liked **Bob Ingram**. He was kind, thoughtful, and always remembered a little girl, bringing me a treat from town, walking around with me and talking to me as though I was important. If we were in town and he saw me, he always bought me an ice cream cone.

Cliff Stauder was in the sheep business and ran sheep by the Mustang. Jeff and my Dad managed his Mustang sheep operation after they came to Colorado from Idaho. Pasturing a bunch of his old ewes in 1931 was the beginning of our flock of sheep. He also owned other land, some south of us, and was in the wool buying business.

John Wilson (and son Johnny) lived near the Red Top. John fixed Mother's sewing machine one time. He was talented at repairs and many hand crafts. They raised horses and mules. When Mother and I visited, I saw many interesting things around their place.

Jim and Tony Spurlock lived by the Cross Bar (near a onetime wagon road route to Pueblo). They ran a few cattle, mostly horses, and seldom rode over to visit us.

Elmer ("Shorty") Baldwin and his wife (children: Sam, Edison, and Juanita) homesteaded east of the Red Top ranch. Their land grew tall, fine grama grass, which they cut and sold to my folks for winter cattle feed.

The Geisingers lived at the head of Myers Canyon just northwest of the Hogback. I only remember them coming by twice.

Mother always smiled when telling about **Vera Jane Pease**, who would come to visit in a surrey with a fringe on top driving a fast trotting little horse. A distinctive character who'd had many pursuits before homesteading, including acting, lecturing, she was well read, dressed colorfully, and was a great talker. Once she met her equal when **Corrine Hart** was also visiting, but she got a head start with her incessant talk, and did not give Vera Jane an opening, leaving Vera frustrated and tapping her foot.

Claude Hart was a neighbor when the three nurses first homesteaded whose land was south of ours. He was the one who found Mother's lost watch on the prairie. Mother later bought part of his land adjoining us to the south, where we now have a good windmill.

The Kenners were a large family (having about ten children) who lived west of North Rattlesnake Butte. They were dry land farmers, and we'd haul milo and other winter stock feed from them. I liked their son Raymond and always hoped he'd be the person to help load. He broke our horse Dixie for $40, but he was not very good at it.

The Erlin Busch family lived this side of North Rattlesnake Butte, and his mother was a Wookey.

Calvin and Earl Mason had a place east of South Rattlesnake Butte.

Dave and Zenith Fosdick (sons Clifford, Frank, and Henry) raised beautiful vegetables and melons in the Arkansas Valley. In summer, we often went to their place to buy their wonderful watermelons and tomatoes. Dave had the gift of water witching for drilling wells, and he often found water where least expected.

Tom Sutton was the mail carrier for many years. His route delivered the mail three days a week to ranchers north of White Rock (such as the Red Top), the Wilsons, and the Mustang area, much of it over wagon road. The mail carrier before him in the early years had only a wagon road to deliver the mail with arroyos to be crossed, a challenging trip in snow and mud. I am not sure where the three nurses picked up their mail, possibly on this route by horseback or maybe in White Rock.

Cowboys

Cowboys were the excitement of my young life whenever one dropped by!

Bill Henderson, the Sanford cowpuncher from 1927, has a eulogy in this book. (page 159)

Joe Eatter, the Red Top buckaroo from their Hardin camp, rode in frequently from the west to visit us during the 1930s, always on a spirited horse, and stayed to enjoy a meal and share news with the folks. He received the normal pay of the time, $30.00 a month. In the 1940s and afterward, the Sandersons and their son Arthur worked at the Red Top and also visited us often.

Shortie Badell, a bachelor cowboy, dropped by on a saddle horse now and then. He later lived in the Geisinger Place near the Hogback. He was proud of his few nice Hereford cattle and some horses.

Tom Hughes lived and worked somewhere around Averson Canyon and used to come in his old car to visit frequently. Often he would ride with me while I was doing cattle work. He enjoyed staying for supper and a few games of cards before heading back.

J. D. Richardson at this time was headquartered at the Petry Place and looking after two of Cliff Stauder's adjoining ranches. The Juan Baca was to the south and the other on the north joined our Mobly pasture. When nearby he rode in frequently to visit. I'd see him come galloping down our lane, cross the arroyo by the windmill, and always stopped there first to give his horse a drink at the tank. His arrival was generally around noontime, and always relished a home-cooked meal with the folks. He especially enjoyed Mother, and after moving to Galt, California, faithfully wrote letters to her.

Endnotes

1 See Pella Historical Society Museums, https://www.pellahistorical.org/copy-of-about-us-1.

2 This family history and many of the details about my mother's early life were provided by her sister, Henriette Brunia Barry, in letters to the author and her mother written in 1974. Information also comes from the book Roorda (n.d. Des Moines, Iowa), containing the genealogy and history of six generations of descendants from Epke Alberts Roorda and Froukje Wijgers vander Lune.

3 Information in this section comes in part from notes taken during conversations with Dad's daughter, Violet Griffith, during one of our visits to Reno in 1951. See also the obituary of William M. Black, Fowler Tribune, August 10, 1951, and "William Black" in James H. Hawley, History of Idaho, vol. 4, although this second source has several mistakes with dates. According to the account of Dad's sister, Mary Ridenbaugh, in History of Idaho, vol. 2, Dad's father had served in the Confederate Army during the Civil War with six of his brothers under General Sterling Price.

4 Vol. 4, p. 35.

5 See "Depression of 1920–21", https://en.wikipedia.org

6 Margaret Black, letter to the author, addressed to Happy Camp, California. Date obscured, probably written before Mother passed away.